Einstein's Gravity and the Expanding Universe: A Cosmic Connection

Jamie

Table of Contents

1   Introduction

1.1   Review of Modern Cosmology

As a data-driven field that studies the origin, evolution, and fate of the Universe as a whole, cosmology has been growing rapidly ever since Einstein proposed a new formulation of gravity (Einstein, 1915), as known as *General Relativity*, and applied it to the Universe (Einstein, 1917). This theory connects the geometry of space-time to the energy-matter content via,

$$G_{\mu\nu} = \frac{8\pi G}{c^4} T_{\mu\nu}, \tag{1.1}$$

where $G_{\mu\nu}$ is the Einstein tensor, $T_{\mu\nu}$ is the energy-momentum tensor, G is the Newtonian constant of gravitation and c is the speed of light in vacuum. Since long before GR, the theory of creation has been the subject of great interest; for instance, Aristotle proposed the idea of the Earth-centered universe in the fourth century, and Newton proposed the static universe in the 17th century, in which he described the Universe to be static, steady-state, and infinite. By the 1920s, Henrietta Swan Leavitt had proved a relationship between the luminosity and periodicity of variable stars, called Cepheids. Her work enabled Edwin Hubble's research to study the velocity-distance relation among extra-galactic nebulae (Leavitt, 1908; Leavitt and Pickering, 1912). Hubble's observations transformed our perspective by proving that not only there are other galaxies outside of our own Milky Way but also that the Universe is expanding (Hubble, 1926, 1929; Hubble and Humason, 1931). Even before these observations, Alexander Friedman & George Lemaitre (Friedman, 1922; Friedmann, 1924; Lemaître, 1927) had solved Einstein's equation finding an unstable static solution. As one of the implications of an expanding universe, Lemaitre suggested the idea of a high-temperature early-stage; a few decades later, Gamow, Alpher, and Herman pioneered the early work on the synthesis of elements in the hot early Universe and predicted a remnant radiation from the early Universe with a temperature around $5K$

(Alpher et al., 1948a; Alpher and Herman, 1948; Alpher et al., 1948b). In 1964, the relic Cosmic Microwave Background (CMB) was accidentally detected by Penzias and Wilson (1965). See Partridge (2007) for a historical review.

A few decades later, following the advancement in the technology of satellites and rockets, precise space-based measurements of CMB temperature made by the Cosmic Background Explorer (COBE) revealed that the CMB has a perfect black body spectrum with $T = 2.7K$, and that there are small fluctuations of the order of 10^{-5}K (Mather et al., 1990; Smoot et al., 1992; Fixsen et al., 1994). These anisotropies are quintessential to our existence, because theoretically speaking, all structures seen today are believed to be the result of the growth of the initial fluctuations such as those in the CMB (Sachs and Wolfe, 1967; Rees and Sciama, 1968). The perfect black body spectrum of the CMB proved that our Universe had started from a dense and hot initial condition, and this observation alone was enough to rule out other competing models like the steady-state theory inspired in the 1940s (Hoyle, 1948; Bondi and Gold, 1948). Steven Weinberg wrote in his book (Winberg, 1972),

"The steady-state model is so attractive that many of its adherents still retain hope that the evidence against it will eventually disappear as observations improve. However, if the cosmic microwave radiation . . . is really black-body radiation, it will be difficult to doubt that the Universe has evolved from a hotter denser early stage."

By the 1990s, type Ia supernovae were established as standard candles which allow measurements of the distance-redshift relationship,

$$m - M = 5 \log_{10}(\frac{d_L(z)}{\mathrm{Mpc}}) + 25, \qquad (1.2)$$

where $m - M$ is the distance modulus and d_L is the luminosity distance, a function of cosmological parameters, measured in Mpc at redshift z. In cosmology, we measure the line-of-sight position of galaxies by measuring the redshift of known emission or absorption features in their spectra. Redshift quantifies the stretch of the photon wavelength in the expanding universe compared to the wavelength of the same line in the rest frame laboratory,

$$z = \frac{\lambda_o}{\lambda_e} - 1,$$ (1.3)

where λ_e and λ_o respectively, refer to the wavelength at the time of emission (i.e., in the rest frame) and the observed (i.e., the stretched) wavelength at the telescope. The observations of distance moduli and redshifts from spectroscopy of type Ia supernovae were conducted by two independent groups, and unveiled that farther away supernovae appeared fainter than they would do in a nonaccelerating universe, implying they are located at a greater distance (Riess et al., 1998; Schmidt et al., 1998; Perlmutter et al., 1999). These observations could be only explained by an accelerating expansion of the recent Universe, requiring a mysterious energy component called *dark energy*. Dark energy is often modeled as a fluid with negative pressure, which is commonly parametrized as (Chevallier and Polarski, 2001; Linder, 2003),

$$w(z) = w_0 + w_a z/(1 + z).$$ (1.4)

In the standard cosmological model, dark energy is described by a cosmological constant, with $w_0 = -1$ and $w_a = 0$.

Dark matter is another unknown in physics, which was initially introduced by Zwicky (1933) to explain the motions of galaxies of the Coma cluster. He argued that a large amount of nonluminous 'dark' matter, much more than the amount of luminous matter that we can observe, should exist to explain the observed extra gravity (Zwicky, 1933, 1937). This type of matter is believed to interact with ordinary matter only through gravitational force. Another direct proof of the existence of dark matter was obtained from a weak lensing study of the Bullet cluster showing the gravitational potential does not follow the luminous matter (Clowe et al., 2006). Massive neutrinos are considered a candidate for dark matter. Specifically, the scenario of neutrino flavor oscillation solved the deficit of the solar and atmospheric neutrino fluxes (Fukuda et al., 1998; Ahmad et al., 2001, 2002), by assuming that neutrino flavors, what we observe at a detector, is a linear superposition of mass eigenstates. These mass eigenstates were once considered dark matter candidates and current constraints on the mass squared difference of neutrino mass-eigenstates have been obtained from studying the neutrino flavor oscillations. Within the framework of the standard model of particle physics, there are three neutrino flavors. Theories with additional non-standard neutrinos, sterile, are under investigation both theoretically and experimentally (see, Palazzo, 2013). Constraining the number of neutrinos is another interesting question in cosmology and particle physics (see, Aartsen et al., 2016). While the ground-based neutrino experiments probe the mass differences between the eigenstates, the cosmological probes such as the CMB and LSS are able to determine the effective number of neutrinos and the sum of neutrino mass, providing a complementary piece of information to complete our understanding of neutrino physics (Lesgourgues and Pastor, 2006; Giunti and Kim, 2007; Patrignani et al., 2016).

The twenty-first century commenced an era of precision cosmology with the release of CMB data from the Wilkinson Microwave Anisotropy Probe (WMAP; Spergel et al., 2003), which probed the high redshift, and large scale structure data of galaxies and quasars from the Sloan Digital Sky Survey (SDSS; Abazajian et al., 2003; Tegmark et al., 2004), which probed the low redshift, recent Universe. A comprehensive analysis of all available data has helped us formulate and shape our theoretical modeling of the dynamics and evolution of the Universe. The current benchmark dark energy model that best-fits all observations, ranging from supernovae data, cosmic microwave background anisotropies to clustering of large scale structure, is the flat Universe with the cosmological constant model, i.e. ΛCDM, which predicts a constant energy density-independent of time. The ΛCDM model can accommodate all of these observations. We found that 32% of the total energy is matter (27% dark matter and 5% ordinary matter), while 68% is explained by dark energy (Akrami et al., 2018; Alam et al., 2020). For a review on observational evidence of hot big bang theory, see Weinberg et al. (2013).

Despite the success of the ΛCDM model, this model suffers a fine-tuning problem and an enormous disagreement between the observed value of dark energy and the theoretical prediction by particle physics. As one of the primary cosmological probes, galaxy surveys aim at constructing clustering statistics of galaxies with which we can investigate the dynamic of the cosmic expansion due to dark energy, test Einstein's theory of gravity, and constrain the total mass of neutrinos and statistical properties of the primordial fluctuations, etc (Peebles, 1973; Kaiser, 1987; Mukhanov et al., 1992; Hamilton, 1998; Eisenstein et al., 1998; Seo and Eisenstein, 2003; Eisenstein, 2005; Sánchez et al., 2008; Dalal et al., 2008). This motivated the design of future ground- and space-based dark energy missions, centered on galaxy surveys, to aim at wide-area surveys (i.e., for greater precision), reaching out to higher redshift (i.e., further back in time and for greater precision).

1.2 Large Scale Structure

The field of large-scale structure cosmology has been substantially advanced by recent torrents of spectroscopic and imaging datasets from ground-based galaxy surveys such as the Sloan Digital Sky Survey (SDSS), Two Degree Field Galaxy Redshift Survey, and WiggleZ Dark Energy Survey (York et al., 2000; Colless et al., 2001; Drinkwater et al., 2010). The SDSS has been gathering data through different phases SDSS-I (2000-2005), SDSS-II (2005-2008), SDSS-III (2008-2014), and SDSS-IV (2014-2019). The clustering of large-scale structure traced by galaxies contains a plethora of information on the initial conditions and time evolution of the energy content of the Universe. Galaxies interact with dark matter only via gravity. The spatial distribution of galaxies follows the underlying gravitational potential field of dark matter, but in a biased way. Different types of galaxies respond differently to the gravitational field of dark matter, and this is parameterized by the bias parameter b,

$$\delta_g = b\delta_{\mathrm{dm}}, \tag{1.5}$$

where δ_g and δ_{cdm} are, respectively, the density contrast of galaxies and that of dark matter (see, Kaiser, 1984). Bigger galaxies tend to occupy bigger halos of dark matter, and likewise smaller and younger galaxies live in smaller halos (see, Zehavi et al., 2005). Common targets of cosmological galaxy surveys include Luminous Red Galaxies (LRGs), Emission Line Galaxies (ELGs), and Quasi-Stellar Objects or Quasars (QSOs). LRGs are relatively old galaxies with a bias around $1.5 - 2$, and have been the target for many galaxy surveys (Tegmark et al., 2006). ELGs are a new class of galaxies with a lower bias but their spectra has a distinct [OII] emission line feature that allows their detection at higher redshift (Geach et al., 2008; Comparat et al., 2013). ELGs are used in a recently completed survey, SDSS-IV eBOSS, and will be targeted for upcoming galaxy surveys such as DESI (DESI Collaboration et al., 2016).

Galaxy survey data are the coordinates of astronomical objects on the sky, i.e., right ascension and declination, with (spectroscopic surveys) or without (photometric surveys) accurate redshift estimates. The redshift can be estimated either by fitting the spectra (spectroscopic) or measuring the brightness and colors of the objects (photometric). While the former yields a more precise estimate for the redshift, the latter provides a faster and more cost-effective redshift estimation. To create a catalogue of objects, images are carefully taken, processed, and analyzed to obtain the coordinates and properties, such as flux and ellipticity of astrophysical sources (see, Lang et al., 2016). Then color-magnitude cuts are applied to each source to identify targets, sometimes for follow-up spectroscopy that returns more accurate redshift estimates. The nominal objective of cosmological galaxy surveys is to probe a bigger area and reach fainter samples of galaxies to create a bigger 3D cosmic volume. Every region of the Universe is a random realization of the underlying true cosmology, therefore, the cosmological information within the data can be only revealed by means of statistical tools. Spatial clustering statistics are then computed using the location of the galaxies to condense and extract information from these catalogues.

1.2.1 Clustering Statistics

We use the spatial clustering statistics to extract information from the spatial distribution of galaxies and quasars. We believe that the density field followed a Gaussian random distribution in the early universe, and for such a Gaussian random field, the two-point (2-pt) statistics carry complete information about the statistical properties of the density field. In the late universe, nonlinear structure formation results in increasingly non-Gaussian components on small scales, introducing nonzero higher-order statistics. Due to the complexity of measuring and modeling theoretical systematics in higher-order statistics, current clustering analyses depend heavily on the 2-pt statistics. In this book, we likewise focus on the 2-pt statistics.

Two point statistics used in cosmology are the power spectrum in Fourier space and the correlation function in configuration s pace. The theoretical m odeling i s s impler in Fourier space than in configuration s pace. This i s due to the expectation that fluctuations at different wavenumbers k grow independently in the linear regime. On the other hand, the measurement is easier in configuration space, because modeling of the survey window effect is complicated in Fourier space, which entangles different modes. In theory, the two estimators are Fourier transformations of each other, but in reality, due to a finite range of data used in each analysis, the two estimators are complementary.

1.2.1.1   Density Contrast

For cosmological analysis of a given survey, it is useful to construct a dimensionless *density contrast*, $\delta(\mathbf{x})$, from the matter/galaxy density field,

$$\delta(\mathbf{x}) = \frac{\rho(\mathbf{x})}{\bar{\rho}(\mathbf{x})} - 1, \tag{1.6}$$

where $\bar{\rho}(\mathbf{x})$ is the expected density of matter/galaxies in position $\mathbf{x}$. In general, it varies across the survey footprint and along the line-of-sight due to redshift failure and completeness. We will discuss more practical ways to construct δ for photometric data in Chapter 2 and for spectroscopic data in Chapter 3.

1.2.1.2   Correlation Function

The correlation function is the 2-pt function of the density contrast field in the configuration space (i.e., as a function of physical separation of two points rather than as a function of Fourier wavemode) is a measure to quantify the level of clustering of the given field δ, and is defined by,

$$\xi(\mathbf{x}_1, \mathbf{x}_2) = \langle \delta(\mathbf{x}_1)\delta(\mathbf{x}_2) \rangle, \tag{1.7}$$

where $\mathbf{x}_1$ and $\mathbf{x}_2$ are the two overdensity locations. Using the assumption of an isotropic and homogeneous universe in the absence of RSD, the correlation function will only depend on the distance,

$$\begin{aligned} \xi(\mathbf{x}_1, \mathbf{x}_2) &= \xi(\mathbf{x}_1 - \mathbf{x}_2), \\ &= \xi(|\mathbf{x}_1 - \mathbf{x}_2|), \\ &= \xi(r). \end{aligned} \tag{1.8}$$

The anisotropic clustering is discussed in Section 1.2.1.4. If there is no clustering beyond random fluctuations in the density field, $\xi(r)$ is zero except at $r = 0$ (shot-noise contributes to $r = 0$). Positive clustering has $\xi(r) > 0$ while negative clustering or anti-clustering is associated with $\xi(r) < 0$. There are various estimators for measuring the correlation function from galaxy surveys (for discussion, see Kerscher et al., 2000), but in this book we will work with the Landy-Szalay estimator which is widely used in cosmology due to its small variance (Landy and Szalay, 1993),

$$\xi = \frac{\langle DD \rangle - 2\langle DR \rangle + \langle RR \rangle}{\langle RR \rangle}, \tag{1.9}$$

where $\langle DD \rangle$, $\langle DR \rangle$ and $\langle RR \rangle$ are, respectively, galaxy-galaxy, galaxy-random, and random-random pair-counts. The random pair-count terms are properly normalized to the number of galaxies. The random particles are distributed based on the survey geometry as a function of location, representing the survey selection mask, and the estimator compares the data pairs to the random pairs to derive the true clustering signal in the presence of various survey effects. Therefore, it is crucial to construct the distribution of randoms (or a selection mask) that reflects the survey systematics, which is one of the goals of my book. See Chapter 3 for more discussion on the catalogue of randoms.

When the data lacks the line of sight information and reliable redshift measurements, which is the case in imaging surveys, we often measure the clustering signal using the angular correlation function as a function of the angular separation θ, that is, the two-dimensional projection of ξ along the line of sight:

$$\omega(\theta) = \langle \delta(\hat{n}_1)\delta(\hat{n}_2) \rangle, \tag{1.10}$$

where $\hat{n}_1$ and $\hat{n}_2$ are the unit vectors pointing at the angular coordinates of each density contrast, separated by θ. In Chapter 2, we use a similar estimator for the pixelated density field rather than individual objects.

1.2.1.3 Power Spectrum

Power spectrum is the two-point function of density contrast in Fourier space, i.e., the conjugate of the correlation function. The relationships between the density field in configuration space and in Fourier space are,

$$\delta(\mathbf{k}) = \int \delta(\mathbf{x})e^{i\mathbf{k}.\mathbf{x}}d^3x \tag{1.11}$$

$$\delta(\mathbf{x}) = \int \delta(\mathbf{k})e^{-i\mathbf{k}.\mathbf{x}}\frac{d^3k}{(2\pi)^3}. \tag{1.12}$$

Equivalently, the power spectrum and correlation function form a Fourier pair,

$$P(k) = \int \xi(r)e^{i\mathbf{k}.\mathbf{r}}d^3r \tag{1.13}$$

$$\xi(r) = \int P(k)e^{-i\mathbf{k}.\mathbf{r}}\frac{d^3k}{(2\pi)^3}. \tag{1.14}$$

Ideally, both statistics should carry identical information (i.e., the linear transformation of one another); in reality, they carry slightly different information because we do not use the full range of separation or wavenumber modes for the analysis. Therefore, we often use both statistics together to exploit the complementary nature (for a discussion, see, Hamilton, 2005).

For imaging data, due to the larger error in the redshift measurement, we often use the 2D angular power spectrum C_ℓ which is a projection of the 3D power spectrum $P(k)$ along the line of sight,

$$C_\ell = 4\pi \int k^2 dk P(k) W_\ell(k), \tag{1.15}$$

with,

$$W_\ell(k) = \int dr\, n(r) j_\ell(kr), \tag{1.16}$$

where $n(r)$ is the true redshift distribution of galaxies and j_ℓ is the spherical Bessel function. C_ℓ is the Fourier conjugate of $\omega(\theta)$,

$$\omega(\theta) = \frac{1}{4\pi} \sum_\ell (2\ell + 1) C_\ell \mathcal{L}_\ell(\cos\theta), \tag{1.17}$$

where $\mathcal{L}_\ell$ is the Legendre polynomials for multipole ℓ. More detailed discussion on the application and estimator of the angular clustering statistics is presented in Chapter 2.

1.2.1.4   Anisotropic Clustering

The observed clustering of galaxies is anisotropic, because of observational realities such as peculiar velocity, redshift uncertainties, or using an inaccurate fiducial cosmology for the redshift to distance conversion. The anisotropies are statistically symmetric along the line of sight, and therefore are parameterized as a function of μ, the cosine of the angle between the wavenumber k, or the separation r, and the line of sight. Therefore, we often describe the anisotropic clustering with correlation function $\xi(r,\mu)$ and power spectrum $P(k,\mu)$. The isotropic clustering can constrain only the term $D_A^2 H^{-1}$; while the anisotropic clustering can allow breaking the degeneracy between D_A and H, and measure each quantity independently (see, e.g., Padmanabhan and White, 2008). From such isotropic or anisotropic analysis, we can obtain dark energy density as a function of redshift from dH/dz and $d^2 D_A/dz^2$, and constrain dark energy parameters, e.g., w_0 and w_a.

In the linear regime, the anisotropic power spectrum is related to the linear power spectrum by (Kaiser, 1987),

$$P(k,\mu) = (1 + \beta\mu^2)^2 P(k), \tag{1.18}$$

where $\beta = f/b$ with the growth rate $f = d \ln D(a)/d \ln a$ [1] (see e.g., Eisenstein, 1997). The anisotropic clustering is often projected into the Legendre polynomials to compress the anisotropic information as power spectrum multipoles:

$$P_\ell(k) = \frac{2\ell + 1}{2} \int_{-1}^{+1} d\mu P(k,\mu) \mathcal{L}_\ell(\mu). \tag{1.19}$$

In Chapter 3 we discuss an estimator of power spectrum multipoles in more detail.

---

[1] D is the growth function and $a = (1 + z)^{-1}$ as the scale factor.

1.2.2 Key Features of Large-Scale Structure

Below we will introduce three crucial features in the two-point clustering statistics of the distribution of galaxies, namely, the correlation function and the power spectrum, that together have become the key science focus of current and future spectroscopic galaxy surveys. The first is Baryon Acoustic Oscillations (BAO). This feature allows a very robust standard ruler test, and therefore provides cosmological distances to various look-back times. The cosmological distances as a function of look-back time depend on the expansion history of the Universe and thus on the dark energy properties. The second feature is redshift-space distortions (RSD), which are distortions in the observed 2-point statistics due to the Doppler effect by the peculiar velocity field of galaxies. The peculiar velocity field follows the gravitational potential of the expanding universe, and therefore it provides ways to constrain the cosmic expansion as well as the nature of gravity. The third feature is the scale-dependent biasing effect caused by primordial non-Gaussianities, which can be utilized to constrain the early Universe.

1.2.2.1 Baryon Acoustic Oscillations

The hot early Universe went through a phase of exponential expansion, called inflation, where the physical size of the Universe expanded by a factor of 10^{26}. We believe that inflation seeded the initial perturbations that grew into all structures observed today, including galaxies. Then, the inflation was followed by a reheating process which increased the temperature of the primordial plasma. In the hot early universe, photons and baryons are in thermal equilibrium due to frequent Thompson scattering of photons off free electrons. The interplay between photon radiation pressure and gravity initiates spherical sound waves in the fluid of baryons[2]-photons at any primordial overdensity peak. Near the epoch of

---

[2]Although from particle physics point-of-view electrons are not considered as baryons, but here we implicitly include electrons in *baryons*.

recombination, as the Universe expands and cools down, photons become incapable of ionizing atoms, and thus protons and electrons combine. The density of free electrons in the baryons-photons plasma drops and photons decouple from baryons, because their scattering rate becomes ineffective comparing to the expansion rate of the Universe. The spherical sound waves that have been propagating then suddenly stall at the epoch of decoupling, and leave their imprint as frozen spherical shells of overdensity in the distribution of photons, baryons and subsequently dark matter. We observe many random superpositions of such spherical shells originated from different primordial overdensity peaks. For more details about the physics of BAO, see Silk (1968); Peebles and Yu (1970); Eisenstein et al. (2007).

The radius of this shell corresponds to the sound horizon scale, the distance traveled by sound waves at the drag epoch (t_d), the moment when photons effectively decouple from baryons (see, Eisenstein and Hu, 1998; Bashinsky and Bertschinger, 2001),

$$
\begin{aligned}
r_d &= \int_0^{t_d} c_s(1+z)dt \\
&= \int_{z_d}^{\infty} \frac{c_s}{H(z)}dz,
\end{aligned}
\tag{1.20}
$$

where z_d and c_s respectively, are the redshift of the drag epoch and sound speed in the early Universe. The sound speed $c_s = c/\sqrt{3(1 + 3\rho_b/4\rho_\gamma)}$ depends on the baryon and photon densities, ρ_b and ρ_γ, and the redshift of the drag epoch depends on the baryon and matter densities. All these quantities are precisely measured from the CMB temperature anisotropies, and therefore we can determine r_d from the CMB data with great precision. The size of this feature is ~ 150 Mpc in today's scale.

Knowing the absolute scale of this feature enables us to conduct the 'standard ruler test'. From galaxy surveys, we observe the BAO feature in the observed coordinates, i.e., angular scales $\Delta\theta$ across the line of sight and the redshift extent Δz along the line of sight. Such observed scales are related to its true physical scales across the line of sight $r_\perp$ and along the line of sight $r_\parallel$ through cosmological distances such as angular diameter distance and Hubble parameter (see, Seo and Eisenstein, 2003):

$$r_\parallel = \frac{c_s \Delta z}{H(z)}, \tag{1.21}$$

$$r_\perp = (1 + z)D_A(z)\Delta\theta, \tag{1.22}$$

where $H(z)$ is the expansion rate at redshift z and D_A is the integrated expansion history defined as,

$$D_A(z) = \frac{c}{1 + z} \int_0^z \frac{dz\prime}{H(z\prime)}. \tag{1.23}$$

Since we can determine the precise physical scale of BAO, i.e., the sound horizon scale r_d, from the CMB data, we can derive $H(z)$ and $D_A(z)$ separately by comparing the physical scale r_d with the observed scales $\Delta\theta$ and Δz:

$$r_d = \frac{c_s \Delta z}{H(z)} \tag{1.24}$$

$$r_d = (1 + z)D_A(z)\Delta\theta. \tag{1.25}$$

This is called the standard ruler test[3]. This feature has been indeed measured from large galaxy surveys, as a single peak at ~ 150 Mpc in the 2-pt correlation function (left panel in Figure 1.1) and a series of harmonic oscillations in the power spectrum in Fourier space (right panel in Figure 1.1) (e.g., Eisenstein et al., 2005; Cole et al., 2005; Anderson et al., 2014b). The constraints from cosmological distances such as $D_A(z)$ and $H(z)$ derived from the BAO measurements have been crucial for probing the dark energy density and time evolution at low redshift (Anderson et al., 2014b).

Due to the large-scale bulk flows and the formation of superclusters and clusters, BAO signals are damped and shifted on small, nonlinear scales at low redshift. It is

---

[3]In case of an unknown size of the feature, we can still assume the spherical symmetry and derive a constraint on the combination of $D_A(z)H(z)$ by using $r_\parallel = r_\perp$ (see, Alcock and Paczyński, 1979). This is called the Alcock Pazynski test, which contains less information than the full standard ruler test.

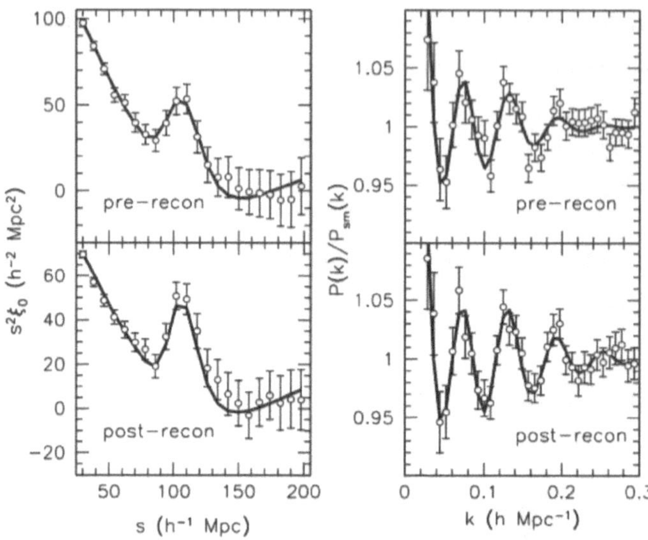

Figure 1.1: Measurements of two-point clustering statistics ($\xi(s)$ in the left panel and $P(k)$ in the right panel) from the CMASS sample from BOSS Data Release 11. In order to highlight the signals in each case, it is common to multiply the correlation function by s^2 and divide the power spectrum by a smooth power spectrum without oscillations ($P_{sm}(k)$). Top row shows the measurements before reconstruction while the bottom row shows the results after the reconstruction. It is noticeable that the reconstruction sharpens the BAO signal. *Source*: Anderson et al. (2014b).

important to correct for such nonlinear effects if we want to derive more accurate and precise measurements. Eisenstein et al. (2007) formulated a recipe to reconstruct the BAO feature from various nonlinearities. Specifically, using the measured galaxy density field and the continuity equation that connects the density field to the displacement field (i.e., the gradient of the displacement is proportional to the density), they showed that we can partially mitigate the nonlinear degradation of the BAO signal, and thus sharpen the peak in the correlation function or equivalently, restoring the higher harmonics of the oscillations in the power spectrum. For example, the bottom panels in Fig. 1.1 show the application of this method to enhance the acoustic peaks in BOSS data by Anderson et al. (2014b).

1.2.2.2   Redshift-Space Distortions

Redshifts serve as the time coordinate as well as the line-of-sight coordinate when studying the expansion rate, such as in Hubble parameter $H(z)$. Another source that can contribute to the observed redshift is the Doppler shift effect due to the motion of galaxies along the line of sight, as known as peculiar velocities. Therefore, when we infer the line-of-sight location of a galaxy based on the observed redshift, assuming that the redshift is purely due to the expansion of the Universe, we are inevitably making an error due to this peculiar velocity component,

$$\mathbf{s} = \mathbf{r} - \frac{\mathbf{r} \cdot \mathbf{v}}{r^2}\mathbf{r}, \tag{1.26}$$

where $\mathbf{r}$ is the true position, called 'real-space' position and $(\mathbf{r} \cdot \mathbf{v})\mathbf{r}/r^2$ is the distortion introduced by the peculiar velocity $\mathbf{v}$ to our measurement $\mathbf{s}$, called 'redshift-space' position. The peculiar velocities of galaxies induce distortions in the clustering pattern in redshift-space; thus they are called redshift-space distortions (RSD; Kaiser, 1987).

RSD is measured statistically in the correlation function or power spectrum of galaxies by comparing the clustering across the line of sight (which is not subject to the Doppler shift effect) and the clustering along the line of sight. Since the peculiar velocities of galaxies are sourced by the gravitational field that is responsible for the structure growth, different gravity models predict a different rate of structure growth and a different amplitude of the peculiar velocity field. We can therefore use RSD to look for discrepancies from the gravity model based on General Relativity.

1.2.2.3 Local Primordial Non-Gaussianity

Understanding the statistical properties of the initial conditions of the Universe is another interesting question in modern cosmology. Single-field inflationary models predict that primordial fluctuations are almost Gaussian and any deviation toward non-Gaussianity is small (Guth and Pi, 1982; Hawking, 1982; Starobinsky, 1982; Bardeen et al., 1983; Acquaviva et al., 2003; Maldacena, 2003a; Zaldarriaga, 2004; Scoccimarro et al., 2004; Creminelli and Zaldarriaga, 2004). However, some alternative inflationary models, which include more fields, generate non-Gaussian perturbations (Linde and Mukhanov, 1997; Maldacena, 2003b; Bernardeau and Uzan, 2002; Lyth et al., 2003; Allen et al., 1987; Kofman and Pogosyan, 1988; Salopek et al., 1989; Chen et al., 2007), see, e.g., Desjacques and Seljak (2010) for a review. The local-type primordial non-Gaussianity is often parameterized as (Matarrese et al., 2000; Verde et al., 2000; Komatsu and Spergel, 2001):

$$\Phi = \phi + f_{\rm NL}(\phi^2 - < \phi^2 >), \tag{1.27}$$

where Φ is the primordial gravitational field, ϕ is a Gaussian field, and $f_{\rm NL}$ is called the nonlinear coupling constant. For a single field inflationary model, we have $f_{\rm NL} = 5(1-n_s)/12 \sim 0.01$, where n_s is the scalar spectral index of primordial power spectrum. Therefore, a non-zero detection of primordial non-Gaussianity, $f_{\rm NL} \gtrsim 1$, will rule out single-field models of inflation and refine our understanding of the early universe (Mukhanov and Chibisov, 1981; Starobinsky, 1982; Hawking, 1982; Guth and Pi, 1982; Allen et al., 1987; Gangui et al., 1993; Falk et al., 1992), see, e.g., Alvarez et al. (2014) for a review.

The current state-of-the-art constraint is $f_{\rm NL} = -0.9 \pm 5.1$ at 68 per cent confidence level from measurements of the bispectrum of cosmic microwave background anisotropies as measured by the *Planck* satellite (Akrami et al., 2019). However, limited by cosmic variance, CMB measurements cannot reach the necessary precision to differentiate between the single-filed and multi-field inflationary models (e.g., Baumann et al., 2009; Abazajian et al., 2016). An alternative route to constrain f_{NL} is to measure the scale-dependent effect it has on the large-scale clustering of biased tracers at lower redshift (Scoccimarro et al., 2004; Dalal et al., 2008; Matarrese and Verde, 2008; Slosar et al., 2008; Taruya et al., 2008; Grossi et al., 2008),

$$\Delta b \propto f_{NL}(b - p)\frac{1}{k^2 T(k)}, \tag{1.28}$$

where $T(k)$ is the transfer function normalized to unity for small wavenumber k, b is the halo bias, and p is a correction factor accounting for the response of the tracer to the halo gravitational field, e.g., 1.6 for recent mergers (Slosar et al., 2008; Reid et al., 2010). Due to the k^{-2} dependence, the biasing effect is most apparent for fluctuations with large wavelengths (see, e.g., Alvarez et al., 2014, for a review). Therefore, constraining primordial non-Gaussianity demands galaxy redshift experiments that survey a huge cosmic volume. Quasars are bright quasi-stellar objects which are detectable from high redshifts thanks to the brightness of their active nuclei, and are consequently the ideal candidate for studying the distribution of matter on high redshifts, and constraining primordial non-Gaussianities with large-scale structure data (Giannantonio et al., 2014; Leistedt et al., 2014; Karagiannis et al., 2014; Agarwal et al., 2014). Using the scale-depend bias effect on galaxy power spectrum measurements, Slosar et al. (2008) obtained $f_{NL} = 8^{+26}_{-37}$ at 68 per cent confidence level from the Sloan Digital Sky Survey (SDSS; Blanton et al., 2017) Data Release 5 (DR5) quasar sample (Adelman-McCarthy et al., 2007). Recently, Castorina et al. (2019) used 148,659 quasars with $0.8 < z < 2.2$ from the SDSS DR14 (Abolfathi et al., 2018) to obtain $-51 < f_{NL} < 21$ at 95 per cent confidence level.

1.3 Challenges

As an upcoming ground-based survey, the Dark Energy Spectroscopic Instrument (DESI) experiment will gather the spectra of thirty million galaxies over 14000 deg^2

starting in late 2021 for five years; this is approximately a factor of ten increase in the number of galaxy spectra compared to those observed in SDSS. This massive amount of spectroscopic data will lead to groundbreaking measurements of cosmological parameters through statistical data analyses of the clustering measurements of the 3D distribution of galaxies and quasars (Aghamousa et al., 2016). Vera Rubin Observatory's Legacy Survey of Space and Time (LSST) is another ground-based survey currently being constructed. It will gather 20 Terabytes of imaging data every night, and will cover 18000 deg^2 of the sky for ten years with a sample size of 2×10^9 galaxies. LSST would provide enough imaging data to address many puzzling astrophysical problems from the nature of dark matter, the growth of structure to our own Milky Way. Given such a data volume, many anticipate that LSST will revolutionize the way astronomers do research and data analysis (Ivezic et al., 2008; LSST Science Collaborations et al., 2017).

The enormously increased data volume provided by DESI and LSST will significantly reduce statistical uncertainties but will require analyses that are more complex and sensitive to unknown systematic effects. A particular area of concern is associated with systematic effects caused by imaging attributes, such as atmospheric conditions, foreground stellar density, and inaccurate photometric calibration. These systematic effects can affect the target galaxy selection and therefore induce non-cosmological fluctuations in the galaxy density field, leading to excess clustering amplitudes, especially on large scales around $k < 0.01\ h\mathrm{Mpc}^{-1}$ (see e.g. Myers et al., 2007; Thomas et al., 2011a,b; Ross et al., 2011; Ross et al., 2012a; Ho et al., 2012; Huterer et al., 2013; Pullen and Hirata, 2013).

To emphasize on the impact of imaging systematics on galaxy clustering, the BOSS LRG samples were thoroughly analyzed by Ross et al. (2011) and Ho et al. (2012) in configuration space and in Fourier space, respectively, finding that stellar density, survey footprint masking, and sky brightness were the primary sources of systematic errors that cause spurious galaxy density fluctuations which contaminates the true cosmological

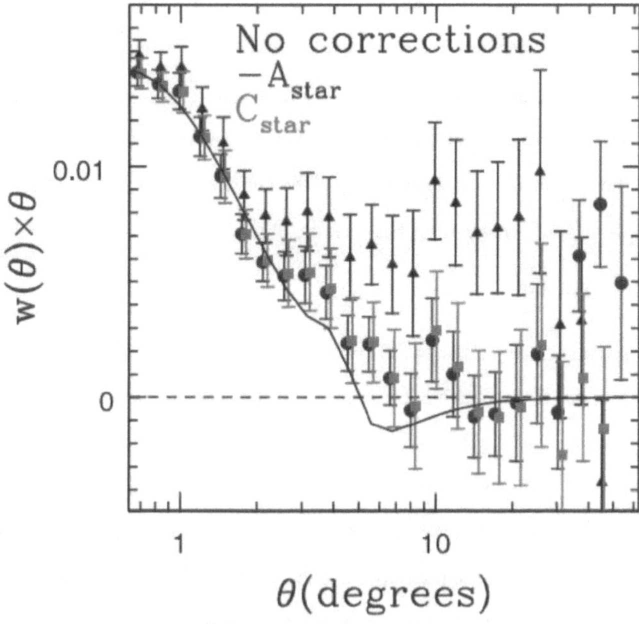

Figure 1.2: Angular clustering of the BOSS Luminous Red Galaxies as a function angular separation before treatment and after correcting for systematics effects. *Source*: Ross et al. (2011).

fluctuations. Taken from Ross et al. (2011), Fig. 1.2 illustrates that the measured angular clustering agrees with the theoretical predication after accounting for correlations against stars ('C_{star}') and area loss due to bright stars ('$-A_{star}$'). Their approach used linear regression, which assumed that the observed galaxy density field is a linear superposition of the systematics and derived the coefficients for each systematic term through computing cross and auto correlations between the observed galaxy density and the systematics.

1.4 Outline

The primary objective of the work presented in this book is to develop and improve techniques from machine learning and data science to clean galaxy survey data, assess the significance of residual systematic error, and unlock robust clustering measurements of large-scale structure. In this book, Chapter 2 introduces a novel methodology based on artificial neural networks and statistics to characterize and mitigate the effect of imaging systematics in a dense photometric galaxy sample from DESI imaging. Chapter 3 enhances the methodology using Poisson statistics for a spectroscopic quasar sample from eBOSS, with a much higher sparsity than DESI galaxies. I also present two statistical tests based on one-dimensional and two-dimensional statistics and realistic simulations to quantify any residual systematic error in real data. Chapter 2 and 3 are papers which I was the first author of, and thus are incorporated into this book with minor modifications. As a second author, I have made pivotal contributions to other projects within DESI and eBOSS, which I briefly discuss in Chapter 5. Finally, Chapter 6 concludes my book with a an outline of my plans for future work.

2 Improving Galaxy Clustering with Deep Learning

Robust measurements of cosmological parameters from galaxy surveys rely on our understanding of systematic effects that impact the observed galaxy density field. In this chapter we present, validate, and implement the idea of adopting the systematics mitigation method of Artificial Neural Networks for modeling the relationship between the target galaxy density field and various observational realities including but not limited to Galactic extinction, seeing, and stellar density. Our method by construction allows a wide class of models and alleviates over-training by performing k-fold cross-validation and dimensionality reduction via backward feature elimination. By permuting the choice of the training, validation, and test sets, we construct a selection mask for the entire footprint. We apply our method on the extended Baryon Oscillation Spectroscopic Survey (eBOSS) Emission Line Galaxies (ELGs) selection from the Dark Energy Camera Legacy Survey (DECaLS) Data Release 7 and show that the spurious large-scale contamination due to imaging systematics can be significantly reduced by up-weighting the observed galaxy density using the selection mask from the neural network and that our method is more effective than the conventional linear and quadratic polynomial functions. We perform extensive analyses on simulated mock datasets with and without systematic effects. Our analyses indicate that our methodology is more robust to overfitting compared to the conventional methods. This method can be utilized in the catalog generation of future spectroscopic galaxy surveys such as eBOSS and Dark Energy Spectroscopic Instrument (DESI) to better mitigate observational systematics.

2.1 Introduction

Many techniques have been developed to mitigate the systematic effects due to imaging attributes such as atmospheric conditions, foreground stellar density, and/or inaccurate calibrations of magnitudes. One can generally classify these methods into the mode projection, regression, and Monte Carlo simulation of fake objects.

The mode-projection based techniques attribute a large variance to the spatial modes that strongly correlate with the potential systematic maps such as imaging attributes, thereby effectively removing those modes from the estimation of power spectrum (see e.g. Rybicki and Press, 1992; Tegmark, 1997; Tegmark et al., 1998; Slosar et al., 2004; Ho et al., 2008; Pullen and Hirata, 2013; Leistedt et al., 2013; Leistedt and Peiris, 2014). In detail, the basic mode projection (Leistedt et al., 2013) produces an unbiased power spectrum and is equivalent to a marginalization over a free amplitude for the contamination produced by a given map. The caveat is that the variance of the estimated clustering increases by projecting out more modes for more imaging attributes. The *extended* mode projection technique (Leistedt and Peiris, 2014) resolves this issue by selecting a subset of the imaging maps using a χ^2 threshold to determine the significance of a potential map. The limitation of the mode-projection based methods is that they are only applicable to the two-dimensional clustering measurements, and they reintroduce a small bias (Elsner et al., 2015). Kalus et al. (2016) extended the idea to the 3D clustering statistics and developed a new step to unbias the measurements (for an application on SDSS-III BOSS data see e.g., Kalus et al., 2018).

The regression-based techniques model the dependency of the galaxy density on the potential systematic fluctuations and estimate the parameters of the proposed function by solving a least-squares problem, or by cross-correlating the galaxy density map with the potential systematic maps (see e.g. Ross et al., 2011; Ross et al., 2012a, 2017a; Ho et al., 2012; Delubac et al., 2016; Prakash et al., 2016; Raichoor et al., 2017; Laurent et al., 2017; Elvin-Poole et al., 2018a; Bautista et al., 2018). The best fit model produces a *selection mask (function)* or a set of *photometric weights* that quantifies the systematic effects in the galaxy density fluctuation induced by the imaging pipeline, survey depth, and other observational attributes. The selection mask is then used to up-weight the observed galaxy density map to mitigate the systematic effects. The regression-based methods often assume a linear model (with linear or quadratic polynomial terms), and use all of the data to estimate the parameters of the given regression model; however, the assumption that the systematic effects are linear might not necessarily hold for strong contamination, e.g., close to the Galactic plane. Ho et al. (2012) analyzed photometric Luminous Red Galaxies in SDSS-III Data Release 8 and showed that the excess clustering due to the stellar contamination on large scales (e.g., roughly greater than twenty degrees) cannot be removed with a linear approximation. Ross et al. (2013) investigated the local non-Gaussianity (f_{NL}) using the BOSS Data Release 9 "CMASS" sample of galaxies (Ahn et al., 2012) and found that a robust cosmological measurement on very large scales is essentially limited by the systematic effects. Their analysis indicated that a more effective systematics correction is preferred relative to the selection mask based on the linear modeling of the stellar density contamination. Recently, Elvin-Poole et al. (2018a) developed a methodology based on χ^2 statistics to rank the imaging maps based on their significance, and derived the selection mask by regressing against the significant maps.

Another promising, yet computationally expensive, approach injects artificial sources into real imaging in order to forward-model the galaxy survey selection mask introduced by real imaging systematics (see e.g. Bergé et al., 2013; Suchyta et al., 2016). Rapid developments of multi-core processors and efficient compilers will pave the path for the application of these methods on big galaxy surveys.

In this chapter, we develop a systematics mitigation method based on artificial neural networks. Our methodology models the galaxy density dependence on observational imaging attributes to construct the selection mask, without making any prior assumption of the linearity of the fitting model. Most importantly, this methodology is less prone to over-training and the resulting removal of the clustering signal by performing k-fold cross-validation (i.e., splitting the data into k number of groups/partitions from which one constructs the training, validation, and the test sets) and dimensionality reduction through backward feature elimination (i.e., removing redundant and irrelevant imaging attributes)(see e.g., Devijver and Kittler, 1982; John et al., 1994; Koller and Sahami, 1996; Kohavi and John, 1997; Ramaswamy et al., 2001; Guyon and Elisseeff, 2003). By permutation of the training, validation, and test sets, the selection mask for the entire footprint is constructed. We apply our method on galaxies in the Legacy Surveys Data Release 7 (DR7) (Dey et al., 2018) that are chosen with the eBOSS-ELG color-magnitude selection criteria (Raichoor et al., 2017) and compare its performance with that of the conventional, linear and quadratic polynomial regression methods. While the effect of mitigation on the data will be estimated qualitatively as well as quantitatively based on cross-correlating the observed galaxy density field and imaging maps, the data does not allow an absolute comparison to the unknown underlying cosmology. We therefore simulate two sets of 100 mock datasets, without and with the systematic effects that mimic those of DR7, apply various mitigation techniques in the same way we treat the real data, and test the resulting clustering signals against the ground truth.

This chapter is organized as follows. Section 2.2 presents the imaging dataset from the Legacy Surveys DR7 used for our analysis. In Section 2.3, we describe our method of Artificial Feed Forward Neural Network in detail as well as the conventional multivariate regression approaches. In this section, we also explain the angular clustering statistics employed to assess the level of systematic effects and the mitigation efficiency. We further describe the procedure of producing the survey mocks with and without simulated contaminations. In Section 2.4, we present the results of mitigation for both DR7 and the mocks. Finally, we conclude with a summary of our findings and a discussion of the benefits of our methodology for future galaxy surveys in Section 2.5.

2.2 Legacy Surveys DR7

We use the seventh release of data from the Legacy Surveys (Dey et al., 2018). The Legacy Surveys are a group of imaging surveys in three optical (r, g, z) and four Wide-field Infrared Survey Explorer (W1, W2, W3, W4; Wright et al. (2010)) passbands that will provide an inference model catalog amassing 14,000 deg^2 of the sky in order to pre-select the targets for the DESI survey (Lang et al., 2016; Aghamousa et al., 2016). Identification and mitigation of the systematic effects in the selection of galaxy samples from this imaging dataset are of vital importance to DESI, as spurious fluctuations in the target density will likely present as fluctuations in the transverse modes of the 3D field and/or changes in the shape of the redshift distribution. Both effects will need to be modeled in order to isolate the cosmological clustering of DESI galaxies. The ground-based surveys that probe the sky in the optical bands are the Beijing-Arizona Sky Survey (BASS) (Zou et al., 2017), DECam Legacy Survey (DECaLS) and Mayall z-band Legacy Survey (MzLS)(see e.g., Dey et al., 2018). Additionally, the Legacy Surveys program takes advantage of another imaging survey, the Dark Energy Survey, for about 1,130 deg^2 of their southern sky footprint (Flaugher, 2005). DR7 is data only from DECaLS, and we refer to this data interchangeably as DECaLS DR7 or DR7 hereafter.

We construct the ELG catalog by adopting the Northern Galactic Cap eBOSS ELG color-magnitude selection criteria from Raichoor et al. (2017) on the DR7 sweep files (Dey et al., 2018) with a few differences in the clean photometry criteria (see Table 2.1). In detail, the original eBOSS ELG selection is based on DR3 while ours is based on DR7. Since the data structure changed from DR3 to DR7, we use `brightstarinblob` instead of `tycho2inblob` to eliminate objects that are near bright stars. In contrast to the original selection criteria, we do not apply the `decam_anymask[grz]=0` cut, as any effect from this cut will be encapsulated by the imaging attributes used in this analysis. Also, we drop the SDSS bright star mask from the criteria, as this mask is essentially replaced by the `brightstarinblob` mask. After constructing the galaxy catalog, we pixelize the galaxies into a HEALPix map (Gorski et al., 2005) with the resolution of 13.7 arcmin ($N_{\mathrm{side}} = 256$) in *ring* ordering format to create the observed galaxy density map.

Table 2.1: The Northern Galactic Cap color-magnitude selection of the eBOSS Emission Line Galaxies (Raichoor et al., 2017). We enforce the same selection for the entire sky. Note that our selection is slightly different from Raichoor et al. (2017) in the clean photometry criteria as explained in the main text.

Criterion	eBOSS ELG
Clean Photometry	0 mag < V < 11.5 mag Tycho2 stars mask
	`BRICK_PRIMARY`==True
	`brightstarinblob`==False
[OII] emitters	$21.825 < g < 22.9$
Redshift range	$-0.068(r\text{-}z) + 0.457 < g\text{-}r < 0.112\,(r\text{-}z) + 0.773$
	$0.637(g\text{-}r) + 0.399 < r\text{-}z < -0.555\,(g\text{-}r) + 1.901$

We consider a total of 18 imaging attributes as potential sources of the systematic error since each of these attributes can affect the completeness and purity with which galaxies can be detected in the imaging data. We produce the HEALPix maps (Gorski et al., 2005) with $N_{\mathrm{side}} = 256$ and oversampling of four[4] for these attributes based on the DR7 ccds-annotated file using the `validationtests` pipeline[5] and the code that uses the methods described in

---

[4] In this context, 'oversampling' means dividing a pixel into sub-pixels in order to derive the given pixelized quantity more accurately. For example, oversampling of four means subdividing each pixel into 4^2 sub-pixels. If the target resolution is $N_{\mathrm{side}} = 256$, the attributes will be derived based on a map with the resolution of 4×256 when oversampling is four.

[5] https://github.com/legacysurvey/legacypipe/tree/master/validationtests

Leistedt et al. (2016). These include three maps of Galactic structure: Galactic extinction (Schlegel et al., 1998), stellar density from Gaia DR2 (Brown et al., 2018), and Galactic neutral atomic hydrogen (HI) column density (Bekhti et al., 2016). We further pixelize quantities associated with the Legacy Surveys observations, including the total depth, mean seeing, mean sky brightness, minimum modified Julian date, and total exposure time in three passbands (r, g, and z). For clarity, we list each attribute below:

- *Galactic extinction* (*EBV*), measured in magnitudes, is the infrared radiation of the dust particles in the Milky Way. We use the SFD map (Schlegel et al., 1998) as the estimator of the E(B-V) reddening. The reddening is the process in which the dust particles in the Galactic plane absorb and scatter the optical light in the infrared. This reddening effect affects the measured brightness of the objects, i.e., the detectability of the targets. We correct the magnitudes of the objects for the Milky Way extinction prior to the galaxy (*target*) selection using the extinction coefficients of 2.165, 3.214, and 1.211 respectively for r, g, and z bands based on Schlafly and Finkbeiner (2011).

- *Galaxy depth* (*depth*) defines the brightness of the faintest detectable galaxy at $5 - \sigma$ confidence, measured in AB magnitudes. The measured depth in the catalogs does not include the effect of Galactic extinction (described above), so we apply the extinction corrections to the depth maps in the same manner.

- *Stellar density* (*nstar*), measured in $\deg^{-2}$, is constructed by pixelization of the Gaia DR2 star catalog (Brown et al., 2018) with the g-magnitude cut of $12 < \mathrm{gmag} < 17$. The stellar foreground affects the galaxy density in two ways. First, the colors of stars overlap with those of galaxies, and consequently stars can be mis-identified as galaxies and included in the sample, which will result in a positive correlation between the stellar and galaxy distribution. Second, the foreground light from stars impacts the ability to detect the galaxies that are behind them, e.g., by directly obscuring their light or by altering the sky background, which will cause a negative correlation between the two distributions. The second effect may reduce the completeness with which galaxies are selected and was the dominant systematic effect on the BOSS galaxies (Ross et al., 2012a). The Gaia-based stellar map is a biased set of the underlying stars that actually impact the data. Assuming that there exists a non-linear mapping between the Gaia stellar map and the truth stellar population, linear models might be insufficient to fully describe the stellar contamination. This motivates the application of non-linear models.

- *Hydrogen atom column density (HI)*, measured in cm^{-2}, is another useful tracer of the Galactic structure, which increases at regions closer to the Milky Way plane. The hydrogen column density map is based on data from the Effelsberg-Bonn HI Survey (EBHIS) and the third revision of the Galactic All-Sky Survey (GASS). EBHIS and GASS have identical angular resolution and sensitivity, and provide a full-sky map of the neutral hydrogen column density (Bekhti et al., 2016). This map provides complementary information to the Galactic extinction and stellar density maps. Hereafter, *lnHI* refers to the natural logarithm of the HI column density.

- *Sky brightness (skymag)* relates to the background level that is estimated and subtracted from the images as part of the photometric processing. It thus alters the depth of the imaging. It is measured in AB $mag/arcsec^2$.

- *Seeing (seeing)* is the full width at half maximum of the point spread function (PSF), i.e., the sharpness of a telescope image, measured in arcseconds. It quantifies the turbulence in the atmosphere at the time of the observation and is sensitive to the optical system of the telescope, e.g., whether or not it is out of focus. Bad seeing conditions can make stars that are point sources appear as extended objects, therefore falsely being selected as galaxies. The seeing in the catalogs is measured in CCD 'pixel'. We use a multiplicative factor of 0.262 to transform the seeing unit to arcseconds.

- *Modified Julian Date (MJD)* is the traditional dating method used by astronomers, measured in days. If a portion of data taken during a specific period is affected by observational conditions during that period, regressing against MJD could mitigate that effect.

- *Exposure time (exptime)* is the length of time, measured in seconds, during which the CCD was exposed to the object light. Longer exposures are needed to observe fainter objects. The Legacy Surveys data is built up from many overlapping images, and we map the total exposure time, per band, in any given area. A longer exposure time thus corresponds to a greater depth, all else being equal.

As part of the process of producing the maps, we determine the fractional CCD coverage per passband, fracgood (f_{pix}), within each pixel with oversampling of four. We define the minimum of f_{pix} in r, g, and z passbands as the *completeness* weight of each pixel,

$$\text{completeness } f_{\mathrm{pix}} = \min(f_{\mathrm{pix,r}}, f_{\mathrm{pix,g}}, f_{\mathrm{pix,z}}). \tag{2.1}$$

We apply the following arbitrary cuts, somewhat motivated by the eBOSS target selection, on the depth and f_{pix} values to eliminate the regions with shallow depth and low pixel completeness due to insufficient available information:

$$depth_r \geq 22.0, \tag{2.2}$$

$$depth_g \geq 21.4,$$

$$depth_z \geq 20.5,$$

$$\text{and } f_{\text{pix}} \geq 0.2,$$

which results in 187257 pixels and an effective total area of 9459 $\deg^2$ after taking f_{pix} into account. We report the mean, 15.9-, and 84.1-th percentiles of the imaging attributes on the masked footprint in Tab. 2.2.

As an exploratory analysis, we use the Pearson correlation coefficient (PCC) to assess the linear correlation between the data attributes. For two variables X and Y, PCC is defined as,

$$\rho_{X,Y} = \frac{cov(X, Y)}{\sqrt{cov(X, X)cov(Y, Y)}}, \tag{2.3}$$

where $cov(X, Y)$ is the covariance between X and Y across all pixels. Fig. 2.1 shows the observed galaxy density after the pixel completeness (i.e., fracgood f_{pix}) correction in the top panel and the correlation (PCC) matrix between the DR7 attributes as well as the galaxy density (*ngal*, the bottom row) in the bottom panel. These statistics indicate that Galactic foregrounds, such as stellar density *nstar*, neural hydrogen column density *lnHI*, and Galactic extinction *EBV*, are moderately anti-correlated with the observed galaxy density. Each of these maps traces the structure of the Milky Way and the anti-correlation with *ngal* implies that, for example, closer to the Galactic plane where the extinction and stellar density are high, there is a systematic decline in the density of galaxies we selected in our sample. The top-left corner of Fig. 2.1 shows that these three imaging attributes are strongly correlated with each other. Likewise, the negative correlation of *ngal* with *seeing* indicates that as *seeing* increases the detection of ELGs becomes more challenging. On the other hand, we find a positive correlation between *ngal* and *depth*s, which can be explained

Figure 2.1: *Top panel*: the pixelated density map of the eBOSS-like ELGs from DR7 after correcting for the completeness of each pixel (see eq., 2.1) and masking based on the survey depth and completeness cuts, see eq., 2.2. The solid red curve represents the Galactic plane. This figure is generated by the code described in https://nbviewer. jupyter.org/github/desihub/desiutil/blob/master/doc/nb/SkyMapExamples.ipynb. *Bottom panel*: the color-coded Pearson correlation matrix between each pair of the DR7 imaging attributes.

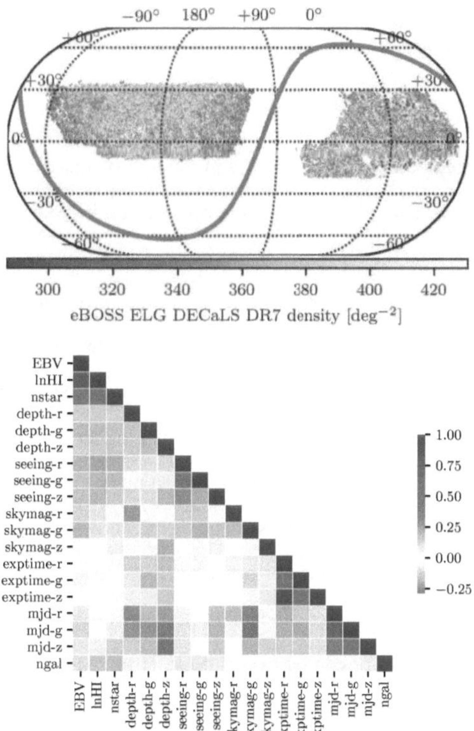

by the fact that as the depth decreases, e.g., we cannot observe fainter objects, the number of galaxies decreases as well.

This matrix overall demonstrates that the correlation among the imaging variables is not negligible. For instance, in addition to the aforementioned correlation among the Galactic attributes, there is an anti-correlation between the MJD and depth values. Likewise, there is an anti-correlation between the seeing and depth values. The complex correlation between the imaging attributes causes degeneracies, and therefore, complicates the modeling of systematic effects, which cannot be ignored and needs careful treatment.

2.3 Methodology

2.3.1 *Observed Galaxy Density*

In our methodology, we treat the mitigation of imaging systematics as a regression problem, in which we aim to model the relationship between the observed galaxy density (*label*) and the imaging attributes (*features*) that are the potential sources of the systematic error. Note that we do not include the positional information as *input* features since we do not want the mitigation to fit the cosmological clustering pattern. The solution of the regression then would provide the predicted mean galaxy density (i.e., in the absence of clustering or shot-noise) solely based on the imaging attributes of that location. We use this predicted galaxy density as the *survey* selection mask to be applied to any observed galaxy map in the attempt to eliminate the systematic effects and therefore isolate the cosmological fluctuation. Below we describe our procedure.

In this chapter, we focus on the multiplicative systematic effects. The observed number of galaxies within pixel i can be expressed in terms of the true number of galaxies n_i and the contamination model $\mathcal{F}(\mathbf{s}_i)$ as

$$n_i^o(\mathbf{s}_i) = n_i \mathcal{F}(\mathbf{s}_i), \tag{2.4}$$

Table 2.2: The statistics of the DR7 imaging attributes used in this chapter. Due to the non-Gaussian nature of the attributes, we report the mean, 15.9-, and 84.1-th percentile points of the imaging attributes.

Imaging map	15.9%	mean	84.1%
EBV [mag]	0.023	0.048	0.075
$\ln(\mathrm{HI/cm}^2)$	46.67	47.21	47.71
depth-r [mag]	23.46	23.96	24.33
depth-g [mag]	23.90	24.34	24.55
depth-z [mag]	22.57	22.93	23.23
seeing-r [arcsec]	1.19	1.41	1.61
seeing-g [argcsec]	1.32	1.56	1.78
seeing-z [arcsec]	1.12	1.31	1.51
skymag-r [mag/arcsec^2]	23.57	23.96	24.39
skymag-g [mag/arcsec^2]	25.06	25.39	25.80
skymag-z [mag/arcsec^2]	21.72	22.04	22.38
exptime-r [sec]	138.8	480.7	551.2
exptime-g [sec]	213.3	680.6	642.2
exptime-z [sec]	261.4	651.6	658.1
mjd-r [day]	56599.3	57232.7	57953.3
mjd-g [day]	56856.3	57358.1	57956.3
mjd-z [day]	56402.4	57005.0	57447.3

where $\mathbf{s}_i$ is a vector representing the imaging attributes $\mathbf{s}$ of pixel i, and the contamination model $\mathcal{F}(\mathbf{s}_i)$ is an unknown function representing the systematic effects which could be either a linear, non-linear, or a more complex combination of the imaging attributes.

Multiplicative systematics are associated with obscuration and area-loss due to foreground stellar density, Galactic extinction, etc. On the other hand, additive systematics are associated mostly with stellar contamination, as described in Myers et al. (2007); Ross et al. (2011); Ho et al. (2012); Prakash et al. (2016); Crocce et al. (2016). When averaged over many pixels, the effect of additive systematics can be absorbed into the constant term of the multiplicative model $\mathcal{F}$, assuming there is no correlation between the imaging maps and the true galaxy density field. The modeling of $\mathcal{F}(\mathbf{s}_i)$ can be approached by a wide variety of techniques, ranging from the traditional methods based on multivariate functions to non-parametric and non-linear models based on machine learning or deep learning such as random tree forests and neural networks (Breiman, 2001; Geurts et al., 2006).

The cosmological information is contained in the true overdensity that is given by

$$\delta_i = n_i/(f_{\mathrm{pix},i}\bar{n}) - 1, \tag{2.5}$$

accounting for the pixel completeness where $\bar{n}$ is the 'true' average number of galaxies. Then,

$$n_i^o = f_{\mathrm{pix},i}\bar{n}\,(1 + \delta_i)\,\mathcal{F}(\mathbf{s}_i). \tag{2.6}$$

This $n_i^o/f_{\mathrm{pix},i}$ is equivalent to the observed *ngal* aforementioned. Since we do not know the true average number density $\bar{n}$ of the data, we estimate $\bar{n}$ from the average of the observed galaxy field,

$$\hat{\bar{n}} = \frac{\sum\limits_i n_i^o}{\sum\limits_i f_{\mathrm{pix},i}}, \tag{2.7}$$

and treat $\hat{\bar{n}} \equiv \bar{n}$. Due to the finite volume of our sample, $\hat{\bar{n}} \neq \bar{n}$ even in the absence of systematic effects. This imposes the well-known integral constraint effect on any clustering analysis. We further ignore any systematic effect on $\hat{\bar{n}}$ due to the fact we use n_i^o; that is, Eq. 2.7 converges to $\bar{n}$ only when $\sum_i f_{\mathrm{pix},i} = \sum_i f_{\mathrm{pix},i}\mathcal{F}_i$. In this sense we are modeling the relative systematic effect without necessarily determining the accurate 'true' $\bar{n}$. We will use simulated results to test our methodology, and the analysis applied to the simulations with a limited footprint will be subject to the similar finite-volume and systematic effects on $\hat{\bar{n}}$, thus providing a fair comparison and means to catch any obvious problem with this approximation.

Finally, we define the normalized galaxy density per pixel t_i,

$$t_i(\mathbf{s}_i) \equiv \frac{n_i^o(\mathbf{s}_i)}{f_{\mathrm{pix},i}\hat{\bar{n}}} = (1 + \delta_i)\,\mathcal{F}(\mathbf{s}_i). \tag{2.8}$$

With this definition, we can estimate the unknown contamination model $\hat{\mathcal{F}}$ (or selection mask) by modeling the dependence of t_i on $\mathbf{s}_i$. When averaged over many spatial positions, the cosmological fluctuations δ_i will be averaged out and therefore the observed t_i averaged across many pixels with the same imaging attribute, should only be a function of $\mathbf{s}$ and return $\mathcal{F}$:

$$< t_i(\mathbf{s}_i) >_i \simeq\, < \mathcal{F}(\mathbf{s}_i) >_i = \mathcal{F}(\mathbf{s}). \tag{2.9}$$

The inverse of the selection mask which is equivalent to the photometric weights (wt^{sys}) in other studies can therefore be used to remove the systematic effects from the observed galaxy number map (cf. Eq. 2.4),

$$\hat{n}_i = \frac{n_i^o}{\hat{\mathcal{F}}} = n_i^o\, wt_i^{\mathrm{sys}}. \tag{2.10}$$

In the following we describe how we obtain $\hat{\mathcal{F}}$ using different regression approaches, e.g., neural networks and multivariate linear functions. From now on, the terms *features* and *label* associated with each data point refer to $\mathbf{s}$ and t of each HEALPix pixel, respectively.

2.3.2   *Mitigation with Neural Networks*

We will apply *fully connected feed forward neural networks* in order to tackle our regression problem. Fig. 2.2 illustrates a schematic diagram of a neuron, the building block of a neural network, which generates the output based on a linear combination of the inputs followed by a nonlinear transformation, the activation function f.

Figure 2.2: A schematic diagram of a single neuron with the activation function f. The neuron takes a set of inputs $\mathbf{x}=(x_1, x_2, ..., x_n)$, multiplies each of them by its associated weight $\mathbf{w}=(w_1, w_2, ..., w_n)$, and sums the weighted values and a thresh-hold (or a constant offset) which is called *bias*, to form a pre-activation value, $z = \sum_{i=1}^{n} x_i w_i + bias$, which is a linear process. The neuron then transforms the pre-activation z to the output using the activation function $f(z)$, which is where the nonlinear process can enter.

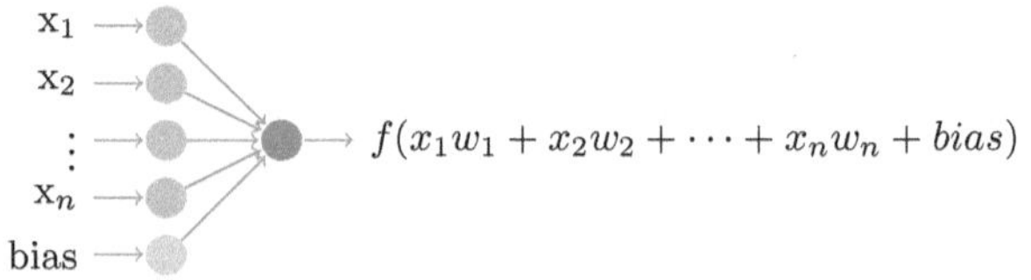

Fig. 2.3 illustrates the architecture of a fully connected feed forward neural network with the imaging attributes in the input layer, three hidden layers of six non-linear neurons in the middle, and a single neuron without any activation function in the output layer, as an example. The *bias* neuron in each layer is shown in orange and is analogous to the intercept in linear regression. The output of the neural network will be an estimation of the contamination model $\hat{\mathcal{F}}$ (see Eq. 2.8). If we use the identity function as the activation function (e.g., $f(z) = z$), regardless of the number of hidden layers, the neural network is equivalent to a linear model. This means that our methodology is a generalization or an extension of the conventional linear mitigation methods. The modeling capabilities of neural networks depend on the number of hidden layers, type of non-linear activation function and the number of neurons in each hidden layer (see e.g. Cybenko, 1989; Hornik

et al., 1989; Funahashi, 1989; Tamura and Tateishi, 1997; Huang, 2003; Lin et al., 2017; Rolnick and Tegmark, 2017).

We use the rectifier $f(z) = \max(0, z)$ as the activation function for the hidden layer neurons, which alleviates the 'vanishing gradient' problem (see e.g., Nair and Hinton, 2010; Glorot et al., 2011; Krizhevsky et al., 2012; Dahl et al., 2013; Montufar et al., 2014).

Figure 2.3: A schematic illustration of a fully connected feed forward neural network with the imaging attributes in the input layer, three hidden layers of six neurons in the middle, and a single neuron on the output layer, as an example. The blue-colored neurons have non-linear activation functions, while the red-colored neuron lacks any activation function. In reality, we employ the validation procedure to choose the best number of hidden layers while keeping the total number of hidden layer neurons fixed at 40 (i.e., approximately twice the number of imaging attributes in this study).

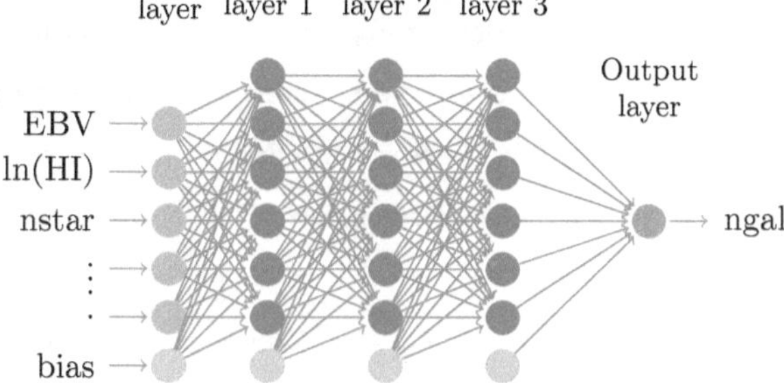

We utilize k-fold cross-validation with $k = 5$ folds/sub-groups to train the parameters, tune the hyper-parameters, and to estimate the predictive performance of the neural network. As illustrated in Fig. 2.4, we randomly split the entire pixel data (187,257 pixels) into five folds and construct the training, validation and test data sets out of these five folds;

three folds are assigned to the training set, one fold is assigned to the validation set, and the remaining one fold is assigned to the test set. A specific assignment of the five folds to these three sets forms one 'partition'. We construct five different partitions such that each fold is used once as test fold. This k-fold cross-validation scheme ensures that a test example is never used for training or tuning.

We standardize (i.e., renormalize) the label and features of the training, validation, and test folds using the mean and standard deviation of the label and features of the corresponding training fold (i.e., similar to Tab. 2.2, but for the training set of each partition). We initialize the biases to zero and sample the weights of each layer randomly from a normal distribution whose variance is inversely proportional to the number of neurons of the previous layer (see e.g., He et al., 2015). Using the training fold, we utilize the adaptive gradient descent with momentum (*Adam*, Kingma and Ba, 2014) to update the parameters of the neural network with batches of size N_{batch}. Thus, the entire training set is split into N_{batch} batches and a gradient update is applied for each batch. One training epoch corresponds to processing the N_{batch} batches once (for more details on the training procedure, we refer the reader to see e.g., Ruder, 2016). The hyper-parameters of *Adam*, specifically the moments decay rates and the tolerance, are fixed as follows: $\beta_1 = 0.9, \beta_2 = 0.999$ and $\epsilon = 10^{-8}$. The default learning rate of 0.001 will be tuned using the validation data.

The network is trained to minimize the following cost function:

$$J = \frac{1}{N_{\text{batch}}} \sum_{i}^{N_{\text{batch}}} f_{\text{pix,i}} \, [t_i - \hat{\mathcal{F}}_i]^2 + \frac{\lambda}{2} \, ||w||^2, \tag{2.11}$$

where the first term is the Mean Squared Error (MSE) weighted with $f_{\text{pix,i}}$, and the second term is the L2 regularization term, used to penalize higher weight magnitudes and a larger number of neurons (Hoerl and Kennard, 1970). The strength of the L2 penalty term is controlled via the regularization scale λ. The network is trained for a number of training

Figure 2.4: A visualization of the five-fold permutations of data-split. The data is randomly split into five equal-size folds, and by permutation of the folds we construct five partitions of data. For each partition/permutation, three folds are assigned to the training set, one fold for the validation set, and the remaining one fold for the test set: therefore, 60% training, 20% validation, and 20% test sets. Each column represents a partition. The test folds are shown in red, while the training and validation folds are shown in blue. The key points are : 1) The folds are not contiguous (in RA, DEC) as may be implied by this cartoon. 2) There is no overlap between the training, validation, and test folds within a partition. 3) One can reconstruct the entire data by merging the test folds from the five partitions.

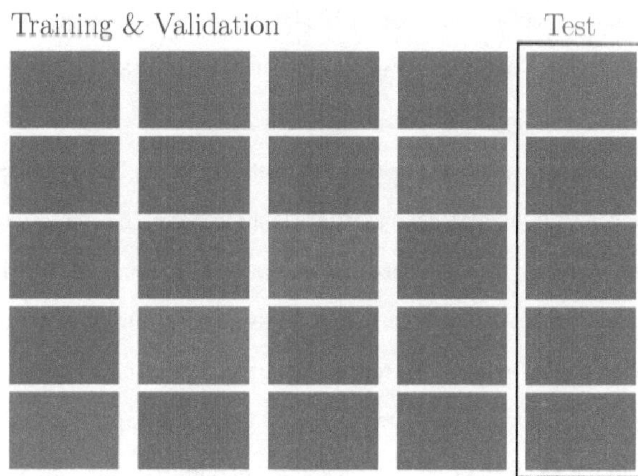

epochs, N_{epoch}, although to avoid unnecessary training, we implement the early stopping technique with the tolerance of 1.e-4 and patience of 10, i.e., the training terminates if the validation MSE does not improve more than the tolerance within the last 10 epochs.

2.3.2.1 Backward Feature Elimination

The input features are highly correlated as shown in Fig. 2.1, and therefore the 18 maps probably contain redundant information. We apply *backward feature elimination* (feature selection) to remove the redundant input features in order to reduce the noise in the prediction as well as to protect the cosmological information by avoiding too much freedom in modeling. We find that reducing the dimension of the input features, i.e., the imaging attributes, is an essential step to avoid over-fitting and regressing out the cosmological clustering.

We perform the feature selection for each partition separately. Initially, we train a linear model on all 18 input features with the following hyper-parameters: the initial learning rate of 0.001, batch-size of 1024, L2 regularization scale of zero, for 500 training epochs with early stopping. We record the validation MSE as a baseline criterion. Then we eliminate one input feature and train the linear model on the remaining 17 features. This trained linear model is applied to the validation set. The input feature whose removal has produced the highest decrease in validation MSE (i.e., the highest improvement in fitting) is permanently eliminated, leaving only 17 features for training. Note that if the feature contained useful information on the systematic effects, removing the feature would have made the fit worse. We repeat the regression using 17 input features, and so on, removing one feature at each iteration until either the validation MSE does not decrease relative to the baseline or all input features are removed. Fig. 2.5 shows the result of the backward feature elimination procedure for the first partition of DR7, ranking the input features based on their importance from left to right. This result supports the trends seen in the exploratory analysis in § 2.2 which indicated strong linear correlations with the stellar density, Galactic extinction, and hydrogen column density in the data (see the correlation matrix in the bottom panel of Fig. 2.1). The color gradient indicates the relative change in validation Root Mean Squared Error (RMSE) when that particular feature is removed with reference to the baseline. We note that the order of removal is not the same as, for example, the color gradient order of the attributes in the first iteration. We believe it is because, as we remove the redundant features, the relevant importance of the remaining input features changes due to the complex correlations between the removed features and the remaining ones. The remaining features for each of the five partitions are shown in Fig. 2.6. The attributes *lnHI* and *nstar* are commonly identified as the most important features and then *ebv, seeing − g, skymag − g, skymag − z, exptime − r, exptime − z, mjd − z* are commonly identified for all 5 partitions.

2.3.2.2 Hyper-parameter Tuning, Training, and Testing

We train the hyper-parameters for each partition separately. At each training epoch and for each choice of hyper-parameters, we apply the trained network on the validation fold. We adjust the hyper-parameters accordingly such that the validation MSE is minimized. Our neural network has the following five hyper-parameters: number of hidden layers; number of training epochs N_{epoch}; L2 regularization scale λ; batch size N_{batch}; Adam's learning rate. We tune one hyper-parameter at a time. To find the optimal learning rate, we monitor the behavior of the cost function during training. We observe that a learning rate of 0.001 leads to a smoothly decreasing cost function vs. training epochs. We train the network for up to $N_{epoch} = 500$ epochs although we implement the early stopping technique with the inertia (or patience) of 10 and the tolerance of 1.e-4: this means the training will be stopped if none of the last 10 epochs achieved a smaller relative error reduction with respect to the minimum validation error, within the tolerance. For the number of hidden layers, we try the following architectures, in which the total number of hidden neurons is fixed at 40 (i.e. roughly twice the number of the features) except for the linear model:

[0] : no hidden layers

Figure 2.5: Feature importance of DR7 based on backward feature elimination for the first partition, as an example. This process iteratively removes the feature whose removal produces the largest decrease in the validation RMSE (i.e., the greatest improvement in fitting) until no decrease is observed. After the first iteration of removal (the top row), the removal of $skymag - r$ decreased the validation RMSE the most and therefore $skymag - r$ is removed. In the second iteration (the second row), removing $exptime - g$ decreased the validation RMSE the most and therefore it is removed. However, in the ninth iteration, the removal of $mjd - z$ did not decrease the validation RMSE and therefore the feature selection stops here, passing the rightmost ten features to the neural network regression. As a result of the process, the importance increases from left to right, and the rightmost ten maps in the figure (ebv, $nstar$, $logHI$, $seeing - g$, $skymag - g$, $skymag - z$, $exptime - r$, $exptime - z$, $mjd - g$, $mjd - z$) are the ones that worsen the validation RMSE when being removed from the input layer.

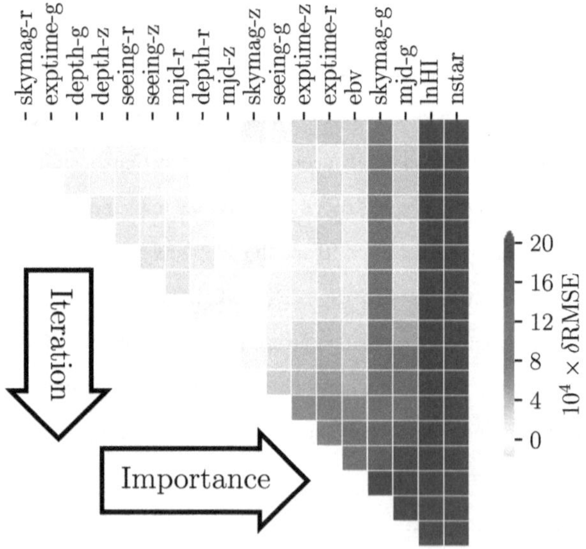

Figure 2.6: Important imaging maps identified by the backward feature elimination (*feature selection*) procedure for the five partitions used for DR7. A darker color of a point within each partition represents a more important attribute identified by the feature selection procedure. Note that *nstar* is identified as the most important attribute in all partitions, i.e., across the footprint.

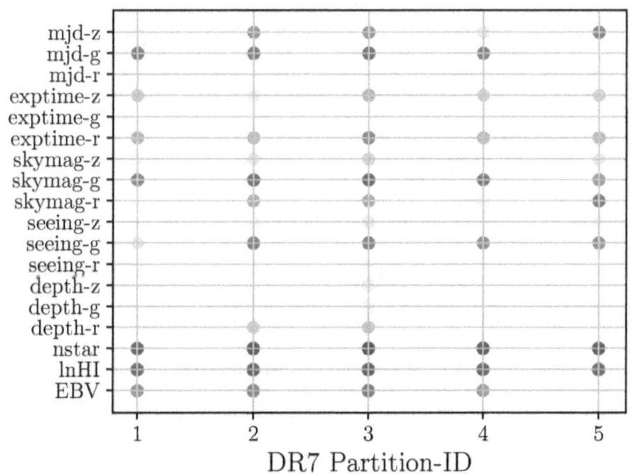

[40] : one hidden layer of 40 neurons

[20, 20] : two hidden layers of 20 neurons on each

[20, 10, 10] : three hidden layers of 20, 10 and 10 neurons

[10, 10, 10, 10] : four hidden layers of 10 neurons

After finding the best number of layers, we proceed to tune λ by trying powers of 10, e.g., 0.001, 0.01, ..., 1, ..., 1000. Finally, we adjust N_{batch} by trying powers of 2, e.g., 128, 256, ... , 4096. The optimal set of the hyper-parameters for each partition is summarized in Tab. 2.3.

Table 2.3: The best hyper-parameters for each partition of DR7.

	number of layers	λ	N_{batch}
Partition 1	[20, 20]	0.001	4096
Partition 2	[20, 10, 10]	0.001	512
Partition 3	[20, 10, 10]	0.001	1024
Partition 4	[20, 10, 10]	0.001	512
Partition 5	[20, 10, 10]	0.001	2048

Once the grid search procedure identifies the best performing hyper-parameters out of the predefined ranges introduced in Section 2.3.2.2, the network is trained with these hyper-parameters for 10 independent runs, each one with a different initialization of the weights and biases, and then applied on the test set. We compute the median of the predicted test label from the 10 runs and aggregate the results over the 5 different partitions to construct the map of the predicted label ($\hat{\mathcal{F}}$) for the entire footprint. For our default method, the backward feature elimination is conducted for each partition and reduces the number of input features before the hyper-parameter training step, as illustrated in Fig. 2.7. This process is performed for each partition separately, each partition using a different fold as the test set, until the entire footprint is covered through the 5 test folds. The flow of the feature selection, hyper-parameter tuning, and testing is summarized in Fig. 2.7.

2.3.3 *Mitigation with Multivariate Linear Functions*

We use linear and quadratic polynomial functions to model the normalized galaxy density dependence on the imaging attributes (Eq. 2.9), as the benchmark approaches to be compared with the neural network. Unlike the proposed neural network method, no regularization or dimensionality reduction is performed, and all data are used to train the parameters of the regression models. Despite a lack of any deliberate machinery against

Figure 2.7: A flow-chart of the backward feature elimination and hyper-parameter tuning for each partition. This entire process is performed five times for each of the five partitions/permutations such that the entire footprint is covered by aggregating the different testing folds.

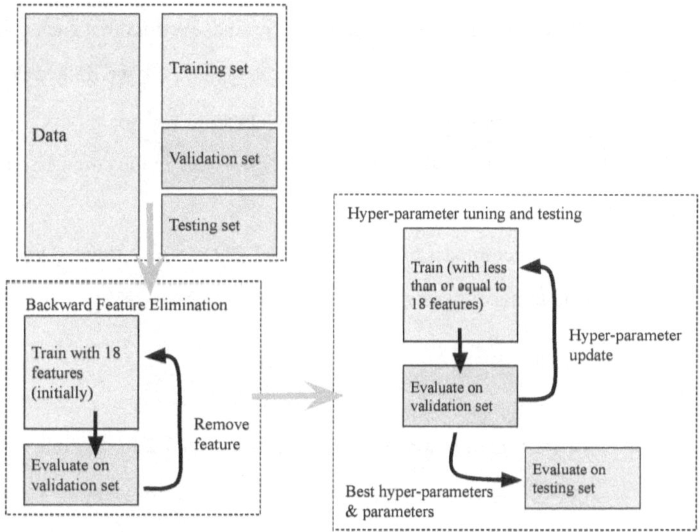

over-fitting, we note that over-fitting is less likely to be an issue for this method since the size of the data is much greater than the number of the fitting parameters. Nevertheless, we tried splitting the sample into 60% of the training data to derive the best fit linear coefficients and 20% of the test set (i.e., the same training and test sample size to have a fair comparison with the neural network) to apply the derived coefficients and permuted five times until the test set covers the entire footprint. We find that such a split does not change the results of the linear regression. On the other hand, the limited flexibility of its parameterized form could be a weakness of this method and we believe it is responsible for the differences that the neural network method makes in the comparison presented in Section 2.4.1.

The contamination model from Eqs. 2.8 and 2.9 can be estimated via a multivariate linear function in terms of the standardized imaging attributes (s) as,

$$\hat{\mathcal{F}}(s|b_0, \alpha) = b_0 + \sum_{m=1}^{M} \sum_{k=1}^{18} \alpha_{mk} (\frac{s_k - \overline{s_k}}{\sigma_k})^m,$$ (2.12)

where M is the maximum power law index equal to 1 and 2 for the linear and quadratic polynomial model, respectively; the constants $\overline{s_k}$ and σ_k are respectively the mean and standard deviation of the k'th imaging quantity, s_k (cf. Tab. 2.2). The parameters b_0 and α_{mk} are the intercept and the corresponding coefficients for each term, respectively, which are tuned by minimizing the weighted sum of the squared errors. The output of the regression is applied as the selection mask on the observed galaxy density to eliminate the systematic effects (see Eq., 2.10).

2.3.4 Angular Clustering Statistics

2.3.4.1 One-point Statistics

In the absence of the systematic effects, the galaxy density field should be statistically independent from the imaging attributes and will only depend on the cosmological fluctuations while an individual dataset/mock will be subject to chance correlations within the statistical error. When averaged over many spatial positions, the cosmological fluctuations will be averaged out and therefore the average density should be equal to the mean density once the survey footprint is accounted for. A deviation from the mean density as a function of the imaging attributes, therefore, is indicative of the average dependence of the observed galaxy density on the corresponding imaging attributes. To assess the level of the contamination in the data, we compute the histogram of the spatially averaged galaxy density vs the imaging attributes. For each of the system attribute, s_k, we prepare 20 bins. For each bin of s_k, we have:

$$\frac{\bar{n}(s_k)}{\bar{n}_{tot}} \equiv \frac{1}{\bar{n}} \frac{\sum\limits_{s_k \leq s_{k,i} < s_k + \Delta s_k} n_i^o}{\sum\limits_{s_k \leq s_{k,i} < s_k + \Delta s_k} f_{\text{pix},i}}, \tag{2.13}$$

where the indices i and k respectively represent the pixel index, and the systematic index; Δs_k is the bin width arranged for different s_k bins such that each bin contains almost the same amount of effective area, in an attempt to suppress fluctuations in $\bar{n}(s_k)/\bar{n}_{tot}$ due to small number statistics. We estimate the error bars using the Jackknife resampling of 20 non-contiguous subsamples of pixels within each imaging attribute bin (see e.g. Ross et al., 2011):

$$\sigma_{\text{Jack}}^2(s_k) = \frac{19}{20} \sum_{j=1}^{20} \left[\frac{\bar{n}(s_k)}{\bar{n}_{tot}} - \frac{\bar{n}_j(s_k)}{\bar{n}_{tot}} \right]^2, \tag{2.14}$$

where $\bar{n}_j(s_k)/\bar{n}_{tot}$ is computed over the entire sample when the j'th Jackknife region is excluded. As a result of the adjusted Δs_k, the level of $\sigma_{\text{Jack}}^2(s_k)$ is almost the same for all s_k. After mitigation of the systematic effects, one expects that the corrected density field is independent of the imaging attributes, i.e., $\bar{n}(s_k)/\bar{n}_{tot}$ being consistent with unity.

2.3.4.2 Two-point Clustering Statistics

The two-point clustering statistic measures the spatial correlation of the galaxy density and has been the main statistic for extracting the cosmological information from galaxy surveys. We use the angular auto and cross two-point clustering statistics of the galaxy density field as well as of the imaging attributes to estimate the impact of the potential systematics on the cosmological clustering signal and to examine the effectiveness of the mitigation techniques tested in this chapter. For pixel i, we calculate the galaxy overdensity δ_i using Eq. 2.5 and the fluctuation of a given imaging attribute δ_i^s as

$$\hat{\delta}_i^s = \frac{s_i}{\hat{\bar{s}}} - 1,\tag{2.15}$$

where $\hat{\bar{s}}$ is the mean of each imaging attribute weighted with $f_{\mathrm{pix},i}$,

$$\hat{\bar{s}} = \frac{\sum_i f_{\mathrm{pix},i} s_i}{\sum_i f_{\mathrm{pix},i}}\tag{2.16}$$

following Ross et al. (2011).

By definition, Eqs. 2.7 and 2.15 ensure that the following integral of the observed quantity over the entire footprint vanishes:

$$\sum_i \hat{\delta}_i \, f_{\mathrm{pix},i} = 0,\tag{2.17}$$

for both the galaxy as well as imaging attribute fluctuations. We utilize both the angular correlation function and angular power spectrum to extract the cosmological information from the galaxy density field. While our mitigation efficiency is evaluated based on the angular power spectrum, we also inspect the angular correlation function to make sure that the systematics are mitigated in that estimator as well, since both estimators are commonly used in the clustering analysis and they are complementary to each other given the limited range of data.

- Angular Correlation Function: we employ the HEALPix-based estimator to compute the angular correlation function which, for a separation angle θ, is defined as (see e.g.

Scranton et al., 2002; Ross et al., 2011)

$$\omega^{p,q}(\theta) = \frac{\sum_{ij} \hat{\delta}_i^p \hat{\delta}_j^q \Theta_{ij}(\theta) f_{\text{pix,i}} f_{\text{pix,j}}}{\sum_{ij} \Theta_{ij}(\theta) f_{\text{pix,i}} f_{\text{pix,j}}}, \tag{2.18}$$

where $p = q$ gives an auto correlation function estimator, $p \neq q$ gives a cross correlation function estimator, and Θ_{ij} is one when two pixels i and j are separated from each other within θ and $\theta + \Delta\theta$, or zero otherwise. Note that our estimator weighs each pixel overdensity with $f_{\text{pix},i}$ since the pixels with a greater complete area coverage should have a higher signal to noise. Such weight is straightforwardly corrected by the denominator in Eq. 2.18 unlike its conjugate estimator (i.e., the power spectrum). Since our overdensity map resolution is limited by the pixel size, we set the $\Delta\theta$ to be the resolution of a pixel (~ 0.23 deg).

- Angular Power Spectrum: one can conveniently expand a coordinate on the surface of a sphere in terms of spherical harmonics or, if azimuthally symmetric, Legendre polynomials. We define the following estimator for expanding the galaxy overdensity:

$$\hat{\delta}_i = \sum_{\ell=0}^{\ell_{\max}} \sum_{m=-\ell}^{\ell} a_{\ell m} Y_{\ell m}(\theta_i, \phi_i), \tag{2.19}$$

where θ, ϕ represent the polar and azimuthal angular coordinates of pixel i, respectively. The cutoff at $\ell = \ell_{\max}$ assumes that the signal power is not significant for modes $\ell > \ell_{\max}$. We define the following spherical harmonic (SH) transform estimator of overdensity ($\hat{\delta}$) over the total number of non-empty pixels N_{pix}:

$$\hat{a}_{\ell m} = \frac{4\pi}{N_{\text{pix}}} \sum_{i=1}^{N_{\text{pix}}} \hat{\delta}_i \, f_{\text{pix,i}} \, Y_{\ell m}^*(\theta_i, \phi_i), \tag{2.20}$$

where * represents the complex conjugate, and we again down-weight the overdensity in pixel i by the completeness ($f_{\text{pix,i}}$). Due to the survey window function implicit in the sum over the non-empty pixels and explicit in $f_{\mathrm{pix,i}}$, our estimator would not return unbiased estimates of the SH coefficients, unless the window function effect is corrected for, and also the expected orthogonalities between different SH modes would not hold. Nevertheless, we define the angular power spectrum estimator as the average of the magnitude of SH coefficients over m:

$$\hat{C}_\ell^{p,q} = \frac{1}{2\ell+1} \sum_{m=-\ell}^{\ell} \hat{a}_{\ell m}^p \hat{a}_{\ell m}^{q*}, \tag{2.21}$$

where $p = q$ gives an auto power spectrum, $p \neq q$ gives a cross power spectrum between the galaxy density and the imaging attributes. In order to compute the angular power spectrum, C_ℓ, we make use of the ANAFAST function from HEALPix (Gorski et al., 2005) with the third order iteration of the quadrature to increase the accuracy[6]. Unlike in the angular correlation function, we do not attempt to correct for the survey window function/survey mask effect in the angular power spectrum both for DR7 and the mocks. We rather calculate the window effect on the theoretical models of power spectrum in Appendix A.1. For the mock test, we use the angular power spectrum observed in the mocks without the contamination model, i.e., the 'Null' case, as our baseline to compare with different mitigation methods.

We use the Jackknife resampling technique with 20 equal-area contiguous regions, as shown in Fig. 2.8, to estimate the error-bars on $\omega(\theta)$ and C_ℓ (see Eq. 2.14).[7] For both the mock and real datasets, we also utilize the cross power spectra between the galaxy

---

[6]We refer the reader to https://healpix.sourceforge.io/pdf/subroutines.pdf, page 104.

[7]We use the mocks without imaging systematics (null mocks in Section 2.3.5) to compare the errors from the Jackknife subsamples of one mock with the errors among 100 full DECaLS-like mock footprints, and we find that the former is greater than the latter on ell $< \sim 10$ by a factor of 2-3, possibly due to the dispersion in the survey window function of the Jackknife samples. For a real survey the observational condition also varies across the footprint. Therefore, with Jackknife errors, the significance of any improvement on systematics treatment will be conservatively assessed.

density and various imaging maps to evaluate the performance of the mitigation. In order to estimate the significance of the contamination in $\hat{C}_\ell^{g,g}$ (or $\omega^{g,g}(\theta)$) before and after mitigation, we calculate $[\hat{C}_\ell^{g,s_k}]^2/\hat{C}_\ell^{s_k,s_k}$ (or $[\omega^{g,s_k}(\theta)]^2/\omega^{s_k,s_k}(\theta)$) as a proxy. [8]

Figure 2.8: Twenty equal-area contiguous regions used to estimate the Jackknife errorbars for the 2D clustering statistics.

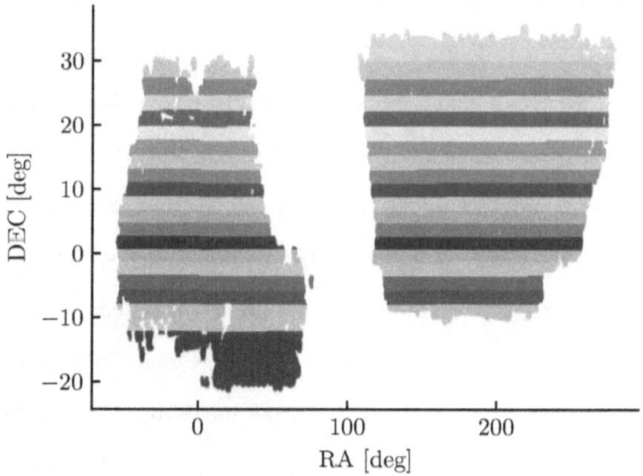

2.3.5 Survey Mocks

Imaging systematics tend to affect the clustering signal mainly on large scales (Myers et al., 2007; Huterer et al., 2013) and the distribution of galaxies on large scales at moderately low redshift can be well-approximated by a log-normal distribution (Coles and Jones, 1991). We therefore believe that log-normal mocks would be sufficient for the purpose of validating our systematic mitigation techniques. We use the *Nbodykit* package

---

[8]These quantities would be the true level of contamination to $\hat{C}^{g,g}$ if the contamination model is linear and systematics are independent of one another (Ross et al., 2012a; Ho et al., 2012).

(Hand et al., 2017) to generate one hundred log-normal cubic mocks with the box-side length of $5274h^{-1}$Mpc and 1024^3 mesh cells, with the input power spectrum matched to the linear power spectrum at $z = 0.85$ based on the *Planck 2015* cosmology (Ade et al., 2016) (i.e., flat ΛCDM with $\Omega_m = 0.3089 \pm 0.0062$, $H_0 = 67.74 \pm 0.46$, $\sigma_8 = 0.8159 \pm 0.0086$), with the galaxy bias of 1.5 and the volume density of $1.947e-4h^3$Mpc^{-3} (see e.g. Raichoor et al., 2017). Then, we use the *make_survey* package (White et al., 2013) to sub-sample the mock galaxies based on the NGC eBOSS ELG redshift distribution in Raichoor et al. (2017) with the redshift cut of $0.55 < z < 1.5$ and to transform the cubic mocks into survey-like mocks. We do not include redshift-space distortions (RSD) in the mocks as we believe that the systematics mitigation efficiency does not depend on the presence of RSD.

The survey mocks are then projected onto the two-dimensional sky using HEALPix and overlaid on the NGC footprint of DR7 to be assigned with the DR7 imaging attributes. Fig. 2.9 illustrates the resulting projection of a simulated survey mock and DR7. Note that the mock footprint (89,672 pixels) is smaller than DR7 (187,257 pixels) almost by a factor of 2. We only use the pixels of the mock that have the DR7 imaging attributes available. The holes (e.g., RA and DEC around 200 and 5 deg) are the pixels that do not have the imaging attributes from the real data. In order to account for the mock survey footprint, we distribute 2,500 random points per deg^2 within the mock footprint and derive the completeness map for the mocks.

2.3.5.1 Null Mocks

Our goal is to develop a systematics treatment methodology that maximally removes the systematic effects while minimally removes the true cosmological signal. The two aspects may not be simultaneously accomplished, in a way that depends on the signal, noise, and the correlation between the imaging maps and the true galaxy density. In our chapter we choose to prioritize losing minimal cosmological information over maximally

Figure 2.9: The projection of the mock footprint (blue) onto the North Galactic Cap of the DR7 footprint (gray). With this projection, the imaging attributes from the real data are assigned to the mocks.

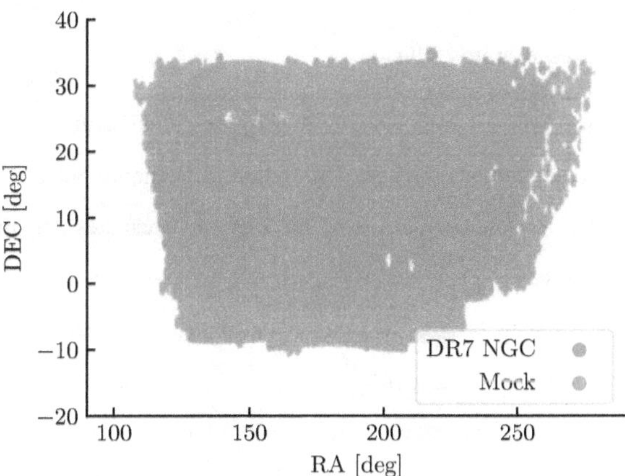

removing systematics. One way of ensuring this is to check if the mitigation method returns the true clustering in the presence of contaminations, which will be tested using the contaminated mocks. Another way is to check if the mitigation method correctly makes a null operation on the clustering in the absence of contaminations, returning $\hat{\mathcal{F}} \simeq 1$. To this end, we utilize the 2D projected mocks without introducing any modulation due to imaging attributes in the galaxy density fields. Henceforth, we call this set of simulations, *null* mocks.

Fig. 2.10 shows the pixel distribution histograms of the number of galaxies per pixel of the mocks in comparison to that of DR7 on the common footprint. The average $ngal = 7.0$/pixel of the null mocks is smaller than $ngal = 13.3$/pixel of the DR7 data (even after accounting for the 5% loss due to tiling completeness and 27% loss due to the redshift range, as stated in Raichoor et al. (2017)). We believe that the difference in *ngal* is due to the different *clean photometry* criteria applied to the ELG selection in Raichoor et al. (2017) and to the targets in this chapter. The standard deviation of *ngal* of the null mocks is 3.0, which is smaller than 4.6 of the DR7 ELGs.

2.3.5.2 Contaminated Mocks

We modulate the mock galaxy density fields using imaging attributes of DR7 and generate the contaminated mocks with additional random noise. The modulation is done based on the best fit coefficients of the imaging attributes and their covariances for $\mathcal{F}$ (Eq. 2.8) that we derived from DR7 using our fiducial linear regression model.

In detail, we pick the 10 imaging attributes, i.e., *EBV*, *nstar*, *lnHI*, *seeing* − *g*, *skymag* − *g*, *skymag* − *z*, *exptime* − *r*, *exptime* − *z*, *mjd* − *g*, *mjd* − *z* of DR7, which were selected from the feature selection procedure on one of the partitions, and modulate the mock density field *n* with $\mathcal{F}$ that is derived from random deviates of the imaging attribute coefficients while accounting for their covariances. Since the measured covariance of such quantities includes both the cosmological fluctuation and the fluctuations due to the imaging attributes, we rescale the measured covariance matrix of the systematics such that the random fluctuation in *ngal* per pixel due to contamination is at a similar level to the cosmological fluctuation from the null mocks. As a result of the random fluctuation in the contamination model we introduced, some of the pixels will be assigned a negative galaxy number. We drop these pixels from our sample. This removes 3.1% of the mock footprint, reducing our mock footprint size from 89,672 to 86,875 pixels. We then introduce the Poisson process, i.e., another random variation step, to ensure the modulated galaxy number per pixel is an integer. These two random variation processes increase the noise in the mock datasets such that the standard deviation of *ngal* of the contaminated mocks (= 4.4) is almost the same as that of DR7, despite the different average *ngal* (see Fig. 2.10).

Figure 2.10: Histogram of the number of galaxies per pixel for DR7 (hatched grey), null (solid blue), and contaminated (dashed red) mocks. All distributions are corrected for the pixel completeness f_{pix}. The DR7 distribution is scaled down to account for 5% tiling completeness, and 27% to 0.55 < reliable redshift ¡1.5. The residual difference between the mocks and DR7 shown here might be due to differences between the *clean photometry* criteria applied to eBOSS target selection and those we apply to DR7.

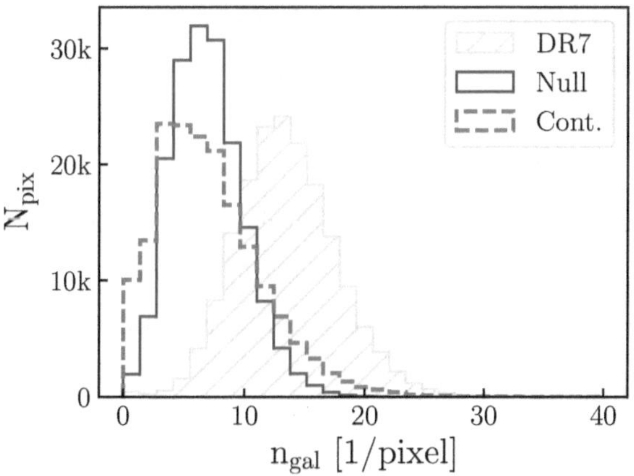

Therefore our mock contamination is simpler than DR7 in that we adopted a linear model, which is chosen purposely since we do not want to give a priori advantage to our neural network method and also since all methods are capable of reproducing the linear model. Meanwhile, this setup is more challenging than the DR7 data since the mitigation is conducted in the presence of a greater level of noise. Note that, while we included only 10 dominant imaging attributes in the contamination, the remaining attributes in the DR7 data are correlated with these 10 attributes and therefore with the modulated galaxy density. All of the mitigation methods in the following mock test will be challenged to deal with such indirect correlations among the 18 attributes. Note that the effect of the footprint, i.e., the survey window effect, is the same for both the null mocks and contaminated mocks since we chose to apply the selection function on the galaxies while leaving the randoms intact. Therefore the null mocks serve as the baseline for estimating the level of systematics in the contaminated mocks.

2.4   Results

In this section, we present the measurements of the clustering statistics before and after correcting for the systematic effects for the real dataset as well as the simulated ones. We demonstrate that the neural network is capable of learning more structure in the observed galaxy density field due to its greater flexibility beyond a fixed functional form, and therefore it can eliminate more excess clustering which is believed to be due to the imaging systematics. We then show the performance of the neural network and multivariate linear models when applied to the mock datasets.

2.4.1   *Mitigating Systematics from DR7*

In the left panel of Fig. 2.11, we show the pixel distribution histograms of the selection masks from the three different regression models we consider in this chapter. While all three models show fairly consistent selection masks for most of the pixels (note the logarithmic scaling of $Npix$), the neural network method (solid red curve) returns extended tails due to a higher representation flexibility associated with its nonlinear nature. We remove pixels with $\hat{\mathcal{F}} < 0.5$ or > 2.0 from our data to avoid too aggressive selection correction since we believe none of these methods can be accurate enough for such a long baseline extrapolation. These pixels account for 1.0% of the original data (from 187,257 to 185,781 pixels). In the right panel, we show the spatial distribution of the removed pixels in the case of the neural network selection mask.

Figure 2.11: The pixel distribution of the selection masks for DR7. *Left*: Distribution of the selection masks (i.e., estimates of the contamination model) derived from different regression models. *Right*: Spatial scatter of the pixels we remove from our data due to the extreme values of the neural network selection mask.

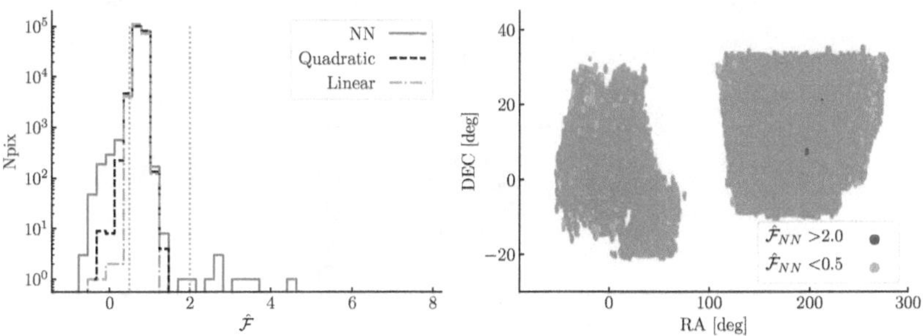

Fig. 2.12 illustrates the spatial distribution of the observed galaxy density before (top left) and after correction (top right) using the neural network selection mask. The bottom panels show the neural network selection mask used for the correction (left) in comparison to the masks derived from the linear (middle) and the quadratic polynomial (right) models. All three masks capture a very similar large scale pattern such as the decrease in the galaxy density close to the Galactic midplane, which is consistent with the negative correlation coefficients between the galaxy density and the Galactic extinction, hydrogen column density, or stellar density. On smaller angular scales, the three selection masks show different fluctuation details. In the following analyses, we examine which method returns the least contaminated galaxy density distribution.

First, once the systematic effects are corrected for, the mean galaxy density should be independent of the imaging attributes. In Fig. 2.13 we show the mean number density of galaxies as a function of the imaging attributes. Again, different bins are set to include

Figure 2.12: *Top*: The normalized observed galaxy density of DR7 before and after systematics treatment using the neural network selection mask, respectively, from left to right. *Bottom*: the selection masks from the Neural Network, quadratic, and linear polynomial models, respectively from left to right. All three selection masks are able to capture the behavior that the galaxy density systematically drops at the footprint boundaries i.e., high extinction regions.

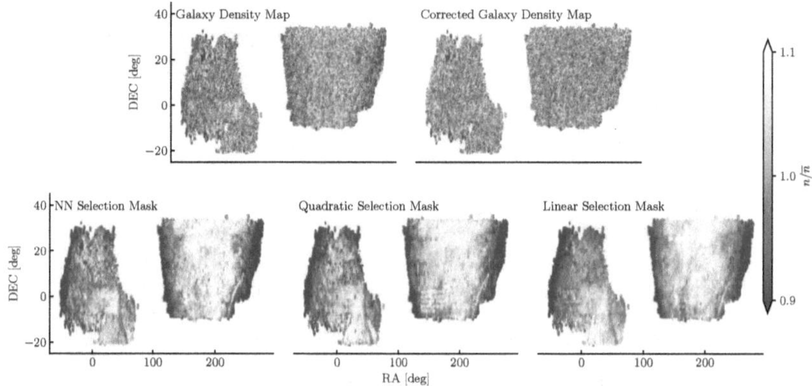

the same effective pixel area and therefore have the same sampling error. The solid black curve shows the galaxy mean density before correction and the solid red shows the result after correction with the selection mask of the neural network model. The dot-dashed curve represents the correction using the linear polynomial model and the dashed curve is for the quadratic polynomial model. The errorbars are computed using 20 Jackknife sub-samples and shown on only one case for clarity. Similar to what we found from the feature selection procedure, the stellar density, Galactic extinction, and HI density exhibit the strongest dependence before correction. After correction, all three methods return the fractional galaxy density close to unity. To quantify the deviation from unity, we report the χ^2 statistics in Tab. 2.4 while ignoring the covariance between the different bins and different imaging attributes. Overall, the neural network achieves the smallest deviation from unity which indicates its highest efficiency in reducing the systematic effects. Ideally we would like to have residual contamination less than the statistical error. Figure 2.13 and Table 2.4 implies that we need to further improve the mitigation techniques for future cosmological analyses. In Section 2.4.3 we provide a more detailed analysis using the same χ^2 statistics and the mocks to quantify the remaining systematics and assess whether or not the data is clean enough.

Table 2.4: The χ^2 values for the measured mean density of the DR7 galaxies vs imaging systematics, presented in Fig. 2.13. This table presents the cumulative values over all bins and all imaging attributes (i.e., N_{bins} =20 bins $\times$ 18 attributes) without accounting for the covariance both between the imaging maps and between different bins[9].

Correction scheme	χ^2	N_{bins}	χ^2/N_{bins}
None	20921.633	360	58.116
Linear	2588.349	360	7.190
Quadratic	2623.006	360	7.286
Neural Network	966.601	360	2.685

We next evaluate the performance of different mitigation techniques using the two-point statistics. We first show the cross power spectra between the DR7 observed galaxy density and various imaging attributes in the form of $[\hat{C}_\ell^{g,s_k}]^2/\hat{C}_\ell^{s_k,s_k}$ in Fig. 2.14. Again, this quantity approximately represents the level of contamination from each attribute to the auto power spectrum of galaxy density and we therefore compare this with the uncertainty in the auto as well as cross power spectrum of galaxies (light and dark gray shades) which

Figure 2.13: The mean number density of the DR7 galaxies as a function of the potential systematics. The solid black curve shows the result before mitigation (*no correction*); the solid red curve is for the result after correcting with the neural network selection mask; the dot-dashed and dashed black curves represent mitigations with the linear and quadratic polynomial selection masks, respectively. The error bars are estimated using the Jackknife resampling of 20 non-contiguous subsamples of pixels within each imaging attribute bin (a total of 20 bins per attribute) and are shown only for one case. This plot again shows that the Galactic foregrounds such as the stellar density introduce a systematic trend in the galaxy density, which indicates a significant contamination by our own galaxy before mitigation. Such systematic trends are mostly removed with any of the three mitigation methods.

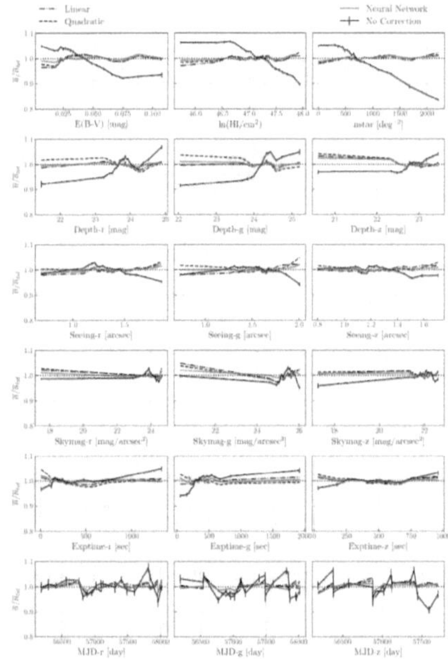

are estimated using the Jackknife resampling of 20 equal-area contiguous regions (see Fig. 2.8). Similarly, we plot $[\omega^{g,sk}(\theta)]^2/\omega^{sk,sk}(\theta)$ in Fig. 2.15 to assess the contamination in the auto-correlation function. Fig. 2.14 and 2.15 show significant contamination on large scales from *ebv*, *lnHI*, and *nstar* compared to the statistical fluctuation estimated from the Jackknife subsampling of the data. The stellar density map shows the highest cross power spectrum with the galaxy density map, which is in agreement with the previous results. Qualitatively, all three mitigation techniques perform well and substantially reduce the cross power below $\ell \sim 30$ and over all separation scales in the cross-correlation function. The neural network method shows a slightly lower cross-power, but this appears to be merely related to the lower amplitude of the corresponding auto galaxy power spectrum compared to the other two cases. We note the spurious peak in the cross-correlation against exptime-z in Fig. 2.15 near the expected angular location of the BAO feature and such feature necessitates thorough investigations of imaging systematics in analyzing the auto clustering statistics of the spectroscopic data for BAO analysis.

We finally present the effect of the imaging attributes before and after mitigation on the auto galaxy clustering statistics. In Fig. 2.16, we illustrate the measured two-point clustering statistics for DR7; the measured angular power spectrum without shot-noise subtraction is shown in the left panel, and the HEALPix-based angular correlation function is shown in the right panel. The solid black curve shows the measured clustering before mitigation, while the corrected measurements using the traditional linear, quadratic polynomial, and the default neural network models are shown respectively with the black dot-dashed, black dashed, and solid red curves. In the right panel, the linear and the quadratic polynomial mitigation results are indistinguishable and overlaid.

---

[9]Note that the best fit neural network model was applied to the unseen data (i.e., the test set) unlike in the linear and quadratic polynomial models. Nevertheless the neural network method returns the smallest χ^2, i.e., the highest efficiency.

Figure 2.14: The cross power spectrum $\hat{C}_\ell^{g,s_k}$ between the DR7 observed galaxy density and the imaging attributes s_k normalized by the auto power spectrum of the imaging attribute $\hat{C}_\ell^{s_k,s_k}$. The plotted quantity $[\hat{C}_\ell^{g,s_k}]^2/\hat{C}_\ell^{s_k,s_k}$ approximately represents the level of contamination to the auto power spectrum of the galaxy density $\hat{C}_\ell^{g,g}$. The light and dark grey shaded regions, respectively, show the Jackknife error estimate of $\hat{C}_\ell^{g,g}$ and $[\hat{C}_\ell^{g,s_k}]^2/\hat{C}_\ell^{s_k,s_k}$ with the galaxy density g before mitigation. The black solid curve shows the result before mitigation (*no correction*), while the solid red curve shows the result after correcting for the systematics with the neural network selection mask. The dot-dashed and dashed black curves show the corrected results with the linear and quadratic polynomial model selection masks, respectively.

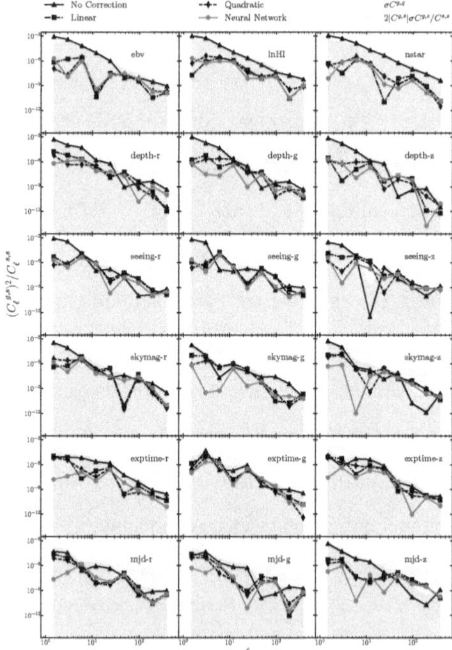

Figure 2.15: The cross correlation function $\omega^{g,s_k}(\theta)$ between the DR7 observed galaxy density and the imaging attributes s_k normalized by the auto correlation function of the imaging attribute $\omega^{s_k,s_k}(\theta)$. The plotted quantity $[\omega^{g,s_k}(\theta)]^2/\omega^{s_k,s_k}(\theta)$ approximately represents the level of contamination to the auto correlation function of the galaxy density $\omega^{g,g}(\theta)$. The grey shaded region shows the Jackknife error estimate of $\omega^{g,g}(\theta)$ before mitigation. All mitigation techniques are able to reduce the excess clustering signal which is due to the imaging systematics.

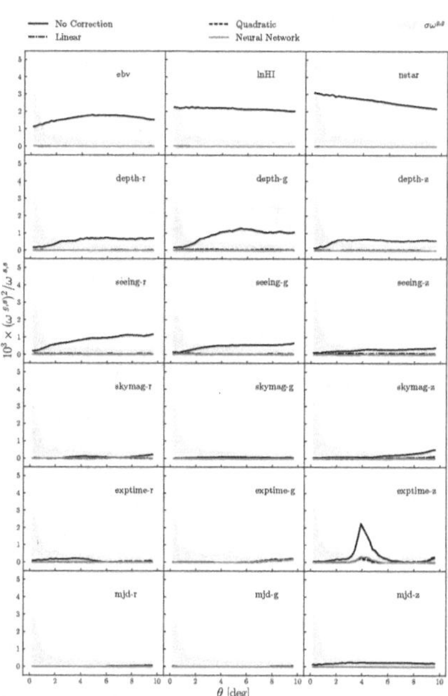

Figure 2.16: Two-point clustering statistics for DR7. *Left*: the measured angular power spectrum without shot-noise subtraction. *Right*: the HEALPix-based angular correlation function. Solid black curves show the measured statistics without correcting for the systematic effects (*no correction*). The dashed and dot-dashed black curves show the statistics after correcting with linear and quadratic polynomial mitigation methods, respectively. The solid red curves show the results after correcting with our default neural network method. The dashed blue curves show the results mitigated with the neural network method but without the feature selection process. The errors are estimated using the Jackknife resampling with 20 contiguous sub-regions and are shown only for a few cases for clarity (see Fig. 2.8).

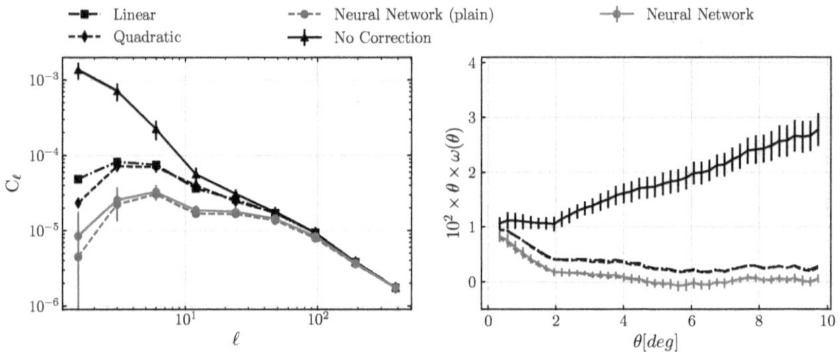

The comparison between the clustering before correction (solid black curves) and after treatment (solid red, blue, dashed, and dot-dashed curves) suggests that the imaging systematics affect the clustering measurements mostly on large scales, e.g., large separation angles or small multipoles, as expected (see e.g., Myers et al., 2007; Ross et al., 2007; Huterer et al., 2013). We find that all of the mitigation methods are able to reduce such large scale contamination, while there still remains substantial excess clustering on large scales, mostly, with the two traditional linear multivariate methods. The neural network method is much more efficient in reducing such excess. When we investigate the effect of the survey window function on this data, we find that the window effect at ℓ ¡ 50 is expected to be less than 5% (more details presented in Appendix A.1, see Fig. A.1).

In comparison to our default neural network model, we also show the measurements mitigated with the neural network model without the feature selection process labeled as 'plain' (blue dashed curves), which is very similar to the default case. In the next section, we test the mitigation methods using the mock datasets for which we know the true clustering signals. As will be demonstrated, our default neural network model with the feature selection process is chosen based on this mock test.

2.4.2 Testing the Mitigation Methods on Mocks

We treat the mocks as if the contamination model was unknown and apply the mitigation pipeline on the mocks as exactly used for the real dataset. After modeling the selection mask for each mock, we remove the pixels whose selection masks values are ¡ 0.5 or ¿ 2.0. This reduces the mock footprint size from 86,875 to 86,867 pixels. Again, the mocks do not include the redshift-space distortions.

2.4.2.1 Feature Selection for Mocks

Fig. 2.17 shows the distribution of the imaging attributes selected by the feature selection process for all of the five partitions of the 100 null (left) and contaminated (right) mocks. For the null mocks, there is no contamination and the feature selection correctly removes most or all of the imaging attributes, as demonstrated by the sparse distribution of the points in the left panel. The imaging attributes that survived feature selection, probably due to a coincidental correlation with the galaxy density, are randomly distributed. On the other hand, the right panel shows that the feature selection procedure correctly identifies most of the input contamination attributes (marked by '*' on the y-axis) for the contaminated mocks and almost always selects *lnHI* and *nstar*. Indeed, as shown in Fig. 2.5, these two attributes were the two most significant input contamination.

Figure 2.17: Important imaging maps identified in the mocks by the feature selection procedure for the five partitions of the 100 null (left) and contaminated mocks (right). The maps used in the input contamination model were marked by '*'. The right panel shows that the feature selection procedure has identified EBV, Stellar density, skymag-g, seeing-g as important in most of the contaminated realizations whereas in the left panel, no map is consistently selected as important among all of the 100 null mocks.

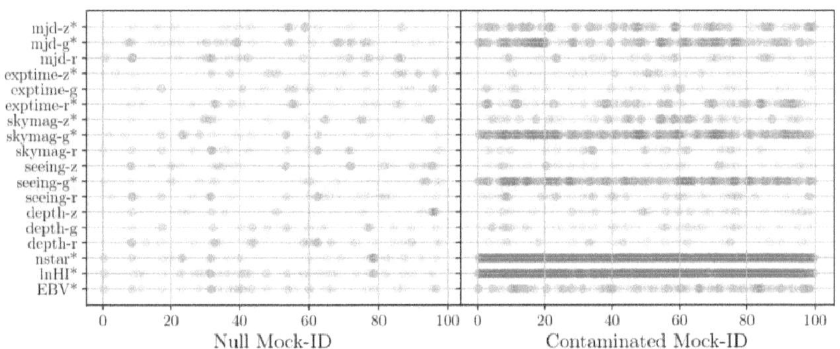

2.4.2.2 Mean Mock Galaxy Density

In Fig. 2.18 we show the number density of mock galaxies, averaged over the 100 mocks, as a function of the imaging attributes. As expected, the galaxy density of the contaminated mocks shows strong or moderate dependence on ebv, $nstar$, $lnHI$, $seeing-g$, $skymag-g$, $skymag-z$, $exptime-r$, $exptime-z$, $mjd-g$, and $mjd-z$ which were indeed the inputs to the contamination model. Meanwhile, the galaxy density also shows strong dependencies on $mjd-r$, $depth-r$, $depth-g$, and $depth-z$ through the correlation between these and the input contamination attributes. Looking at this result alone from a real data perspective, one would not be able to single out the underlying imaging attributes that are directly responsible for the contamination. When the inputs to the mitigation procedure include all of the input contamination maps, Fig. 2.18 shows that all methods effectively remove the dependence. In subsection 2.4.3, we discuss further how well the underlying true mean density is recovered after mitigation.

2.4.2.3 Angular Power Spectrum of Mock Galaxies

Fig. 2.19 shows the mean angular power spectrum of the 100 null and 100 contaminated mocks in the left and right panel, respectively, in the top row. In the middle row, we illustrate the remaining bias[10] on clustering after mitigation as an offset from the true power spectrum (i.e., the null power spectrum from the left panel[11]). One can see that the contamination substantially increased power at $\ell < 50$. Since the contamination model is based on the linear polynomial model, all three fitting methods, i.e., the linear, quadratic, and neural network, are capable of reproducing the true input contamination, while they perform differently in the presence of the two layers of noise we added and the eight additional imaging attributes that are non-trivially correlated with the ten input contamination attributes.

---

[10]Due to the two-step noise introduced during contamination, the shot noise of the contaminated power spectra is increased by almost a factor of two. We estimate the total offset noise to be around 3.05×10^{-6} from $\sigma^2(ngal)/\bar{n}^2$ and subtract it from the power spectrum of the contaminated mocks. The additional shot noise is mostly originated from the Poisson process we applied, i.e., precisely $1/\bar{n}$, where $\bar{n}$ is the average number density after contamination, while an extra $\sim 10\%$ is also due to the noise we added to the contamination model.

[11]Note that we use the clustering of the null mocks before mitigation as the ground truth model since the survey window for all of the mocks before and after systematics treatment does not change.

Figure 2.18: The number density of the mock galaxies as a function of the potential systematics averaged over the 100 mock datasets. The grey shaded region illustrates $1-\sigma$ dispersion in the null mocks. The dotted curve shows the mean density of the null mocks. The black solid curve shows the mean density dependence on the imaging attributes for the contaminated mocks. The solid red curve shows the mean density after correcting for the systematics with the neural network selection mask. The dashed black curve shows the corrected results with the quadratic polynomial model selection mask. The result of the linear polynomial model is almost unity, and therefore is omitted for clarity.

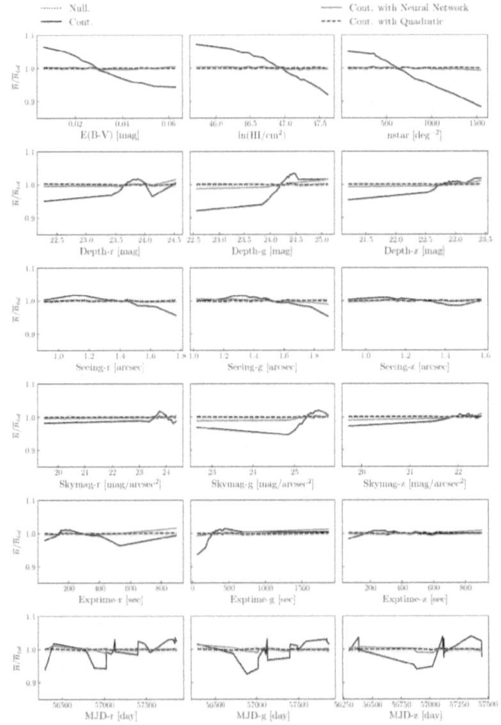

Figure 2.19: *Top row*: The mean angular power spectrum of the 100 contaminated (null) mocks in the right (left) panel. *Middle row*: The mean power spectrum subtracted by the mean of the null mocks to better visualize the remaining bias after each mitigation. The dark grey shaded region shows the $1-\sigma$ confidence region of the mean of 100 mocks, while the light grey area shows the $1-\sigma$ confidence region for one mock, calculated from the dispersion of 100 mocks. To account for the increased shot noise during contamination, we remove the same constant power from all contaminated/mitigated power spectra until their small scale power matches that of the null mock power spectrum. We quantify the significance of the remaining bias by calculating χ^2, the sum of the squared offset weighted with the diagonal variance of the mean C_ℓ of null mocks over the last six bins ($\ell \geq 12$). The middle panel on the left illustrates the neural network without feature selection ('plain') tends to remove the cosmological clustering signal.

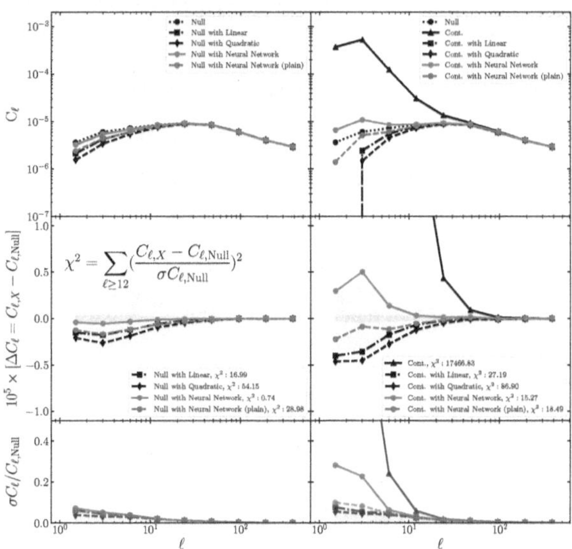

In the right top and middle panel, we find all three methods effectively remove the contamination over $\ell < 100$; in detail, the linear (black dot-dashed) and the quadratic polynomial (black dashed) methods slightly over-correct power while the neural network method (solid red) slightly under-corrects it. Note that, without the feature selection process (dashed blue), the neural network method would also over-correct the large scale power like the linear and quadratic polynomial models. The left top and middle panel show that, in the absence of contamination, both the linear and quadratic polynomial methods over-correct the large scale power since the fitting methods can always find purely coincidental consistency between the imaging attributes and the cosmic variance. The quadratic method that has a greater freedom appears more prone to such problem. On the other hand, the neural network method without the feature selection process shows a lesser degree of overfitting than the linear methods, probably due to the validation procedure. Our default neural network method, which incorporates feature selection, is the most robust mitigation methodology against overfitting.

The remaining bias can be compared to the typical error expected for such data. The dark and light grey shaded regions in the middle panels indicate the 1-σ confidence regions for the mean and the individual mock of the 100 mocks, respectively. We quantify the significance of such remaining bias by calculating χ^2, the sum of the squared offset weighted with the diagonal variance for the mean at each ℓ bin. Note that we use the variance from the 100 null mocks for calculating χ^2 of the contaminated mocks in order to avoid an advantage of the increased variance after contamination. We find that the default neural network with $\chi^2 = 0.74$ (reduced $\chi^2 = 0.12$ with $dof = 6$) recovers the true underlying clustering when applied to the null mocks well within 1σ C.L. of the sample variance. We estimate the significance with taking the residual systematics as one extra degree of freedom,

$$\text{max systematics} \sim \sqrt{\chi^2} \tag{2.22}$$

Figure 2.20: Dependence of χ^2 on the lowest bin $\ell_{\min}$ in the null (left) and contaminated (right) mocks. To better quantify the residual bias introduced by each method, we evaluate the dependence of the bias on the lowest bin $\ell_{\min}$ that is included in the χ^2 computation. The default neural network method performs significantly better than the conventional methods for the null mocks, mainly because the feature selection procedure successfully prevents the method from regressing out the cosmological clustering signal. For the contaminated mocks, all methods tend to perform similarly, as expected, since all mitigation methods can reproduce the input contamination model.

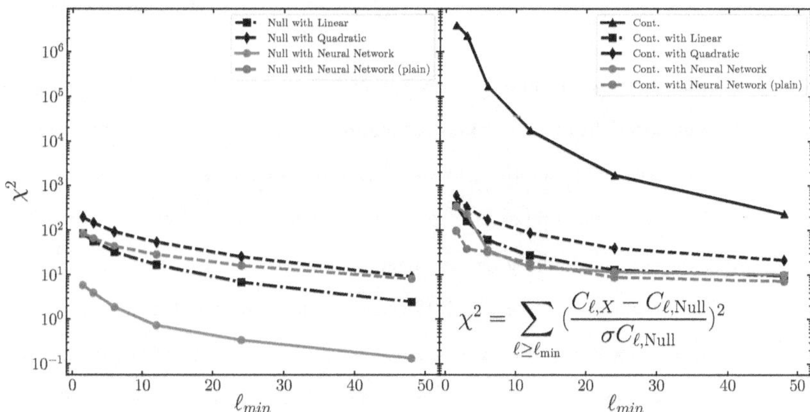

$$\chi^2 = \sum_{\ell \geq \ell_{\min}} \left(\frac{C_{\ell,X} - C_{\ell,\mathrm{Null}}}{\sigma C_{\ell,\mathrm{Null}}} \right)^2$$

On the other hand, the linear and quadratic models have systematic biases with more than 4σ and 7.4σ significance. When applied to the contaminated mocks, we find that the neural network method returns $\chi^2 = 15.3$ from six ℓ bins ($\ell \geq 12$), while the linear and quadratic methods, respectively, polynomial mitigation return 27.2 and 86.9. The difference in χ^2 among the different mitigation methods is not significant compared to the $\chi^2 = 17,466.8$

before mitigation. While the neural network model appears to perform the best, $\chi2$ of 15.3 (reduced χ^2 of 2.6) indicates that the residual is much more significant than the sample variance. However, for the contaminated mocks, we added substantial statistical noises, almost doubling the noise. If we use the covariance of each contaminated/mitigated case, we get χ^2 of 7.7 (reduced $\chi2$ of 1.3) for our default case; that is our residual is at the level of the statistical noise we added in the process of contamination. These χ^2 values can be translated to $2.8 - 3.9\sigma$ uncertainties in the residual systematics (see Eq. 2.22) which implies any particular cosmological study requires a more thorough analysis of systematics to determine an estimate of the residual systematic uncertainty for the parameters of interest.

Such χ^2 can depend on the lower limit of ℓ we consider. In Fig. 2.20, we investigate the behavior of the remaining bias depending on ℓ_{min}, which shows that the neural network method consistently returns a lower remaining bias for various $\ell_{\min}$ choices. However, for the contaminated mocks, the difference is small and we conclude that all mitigation methods perform similarly for the contaminated mocks.

The bottom panels of Fig. 2.19 compare the noise introduced by the three different mitigation processes. The right panel shows that the contamination process (solid black) introduces additional noise on large scales relative to the variance of the null mocks (the gray shade). After mitigation, the variance is reduced, which is probably related to the decrease in the large scale power, since the variance of power is proportional to the amplitude of power itself in the Gaussian limit. The quadratic method shows the smallest fractional error on large scales due to the reduced amplitude after correction. In all cases, if we calculate the fractional variance with respect to the measured C_ℓ (e.g., $\sigma C_{\ell,NN}/C_{\ell,NN}$ instead of $\sigma C_{\ell,NN}/C_{\ell,Null}$), it agrees with the fractional variance of the null mock (gray shade). Therefore, we do not observe a nontrivial increase in variance by any of the mitigation methods we tested.

2.4.2.4 Cross Power Spectrum of Mocks and Imaging Properties

In Fig. 2.21, we show the mean cross power spectrum of the 100 mock catalogs and the imaging attributes for the different mitigation techniques. All three methods substantially reduce the cross power with the imaging attributes. The neural network method tends to show a small residual for $\ell < 10$ that is greater than those of the linear and quadratic polynomial models. The dark shaded region shows the $1-\sigma$ confidence region of the mean cross power propagated to $C^2_{s,g}/C_{s,s}$ and the light shaded region shows the 1σ confidence interval of the mean auto-power spectrum of the mocks as shown in Fig. 2.14. Therefore, we find that these residuals are greater than the statistical noise of $C^2_{s,g}/C_{s,s}$, but the effect on the *auto power spectrum* are marginal for $\ell > 10$. This excess on small ℓ is partly due to the greater auto galaxy power spectrum amplitude (in Fig. 2.19) after mitigation than those by the other methods. Meanwhile we still find a residual correlation with skymag-z and mjd-z even after accounting for the effect of the auto power spectrum amplitude. Without the feature selection procedure (dashed blue), the neural network also returns a smaller residual. In essence, we see that once feature selection is applied, the neural network only corrects to a certain level, controlled by the specifics of the feature selection procedure. This protects against over-fitting due to random correlations between the imaging attributes and the galaxy density field.

As a sanity check, if we use the true input linear contamination model to mitigate the systematics (purple dot-dashed line in Fig. 2.21), the cross-correlation completely vanishes as expected[12]. For the null mocks, all mitigation methods return negligible cross-correlation, which is omitted from Fig. 2.21 for clarity.

Overall, while the cross-correlation statistics between the galaxy density and the systematics attributes are a useful indicator for the level of contamination, we find it may

---

[12]The auto-power spectrum of the contaminated mocks mitigated with the true input contamination model (in Fig. 2.22) returns the smallest residual bias relative to the uncontaminated clustering, as expected.

Figure 2.21: The mean cross power spectrum of the contaminated mock catalogs and the imaging attributes for different mitigation techniques. Our default neural network method with feature selection is shown in solid red while the performance without feature selection ('Neural Network (Plain)') is shown in dashed blue. The dark grey shaded region shows the 1σ confidence region of the plotted quantity derived from $2\sigma(C_{g,s}) * C_{g,s}/C_{s,s}$, and the light grey region shows the typical 1-σ confidence region of the mean auto power spectrum of the 100 mocks. The mitigation with the ground truth contamination model is shown with a purple dot-dashed curve as 'cont. with linear (truth)'.

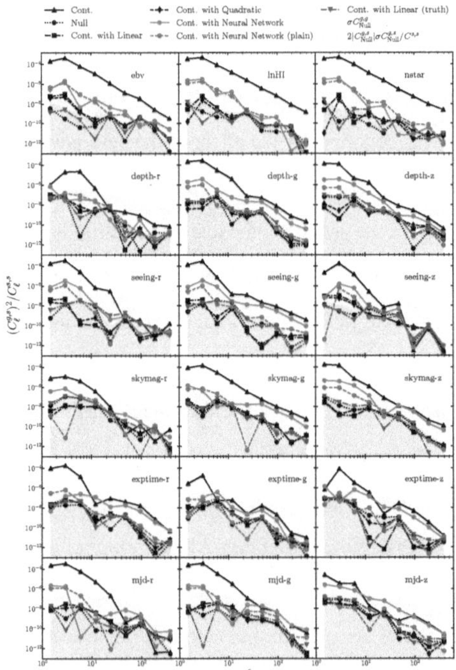

be difficult to infer and discriminate the level of contamination in the density field from such cross-correlation statistics beyond what can be probed by the auto power spectrum.

2.4.2.5 A Case with Underfitting

It is possible that we may identify only a subset of the contamination attributes for a given data set and attempt to mitigate the contamination based on such limited information. We consider a situation where we input only five imaging attributes to the mitigation procedure: four from the true contamination inputs, i.e., EBV, $lnHI$, $nstar$, $skymag - g$ and one that is not among the true contamination inputs, but correlated with the contamination inputs, i.e., $depth - r$. The neural network method could be more resilient to such limited information since its nonlinear activation function may allow the mitigation procedure to better utilize the correlation between different input imaging attributes. In Fig. 2.22, the linear polynomial method with 'few' inputs shows a lesser degree of overfitting for the null mocks, compared to the default linear case, while showing under-fitting for the contaminated mocks. This is expected as the freedom of the linear model is now limited. The neural network method with the 'few' inputs (without feature selection) returns a very similar pattern as the linear 'few' case, which implies that our current default neural network method does not have an advantage over the linear model in such a case despite its greater flexibility. This underfitting case may be worth future investigation, as apparent excess clustering remains in DR7 in Fig. 2.16 when any method is applied, but especially in the case of the application of multivariate linear models.

2.4.3 *Summary and Discussion*

In summary, we find that our default neural network method is more robust against overfitting based on the test with the null mocks. This is due to the feature selection process that appropriately reduces the flexibility of mitigation. Based on the tests with the contaminated mocks, we find that both the linear and neural network methods perform

Figure 2.22: Same as Fig. 2.19 showing the mean auto power spectrum of the 100 null (left) and contaminated (right) mocks mitigated with the fewer imaging maps 'few' (to demonstrate under-correction), neural network without the feature selection 'plain', and the ground truth contamination model 'truth'. The mitigation with the ground truth model achieves the lowest residual bias as expected. For the null mocks, using fewer imaging maps prevents over-fitting simply by providing less freedom in the regression model while it leads to underfitting for the contaminated mocks.

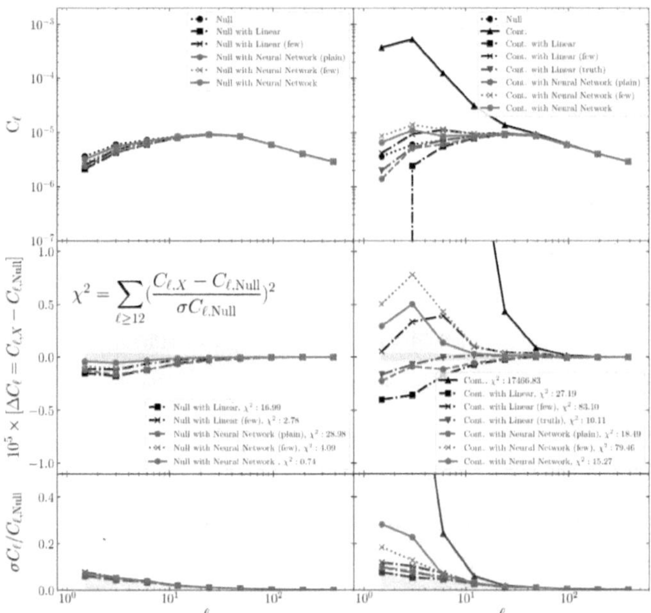

$$\chi^2 = \sum_{\ell \geq 12} (\frac{C_{\ell,X} - C_{\ell,\mathrm{Null}}}{\sigma C_{\ell,\mathrm{Null}}})^2$$

equally well in terms of the residual bias, while the neural network method is more robust against overfitting. The quadratic polynomial method appears to be more prone to the overfitting problem than the other two methods since it has a greater flexibility than the input contamination model, but without a way to suppress the flexibility. All methods do not increase fractional variance during the mitigation process. Note again that we deliberately choose the linear model in contaminating the mocks in this test in order to prevent a disadvantage in using the linear and quadratic polynomial methods. Therefore, the decent performance of the linear method is warranted. In real data, the contamination due to observational effects can take a more complex form as implied by the difference in the mitigation results between the data and our mocks. Therefore our mock test is a conservative estimation of the comparative mitigation capability of our default neural network method.

While we demonstrated qualitatively and quantitatively that our fiducial neural network method is more robust than the conventional methods, for both DR7 as well as for the mocks, Fig. 2.13-2.14 show non-negligible residual contamination compared to the expectations. It is not surprising that the mitigation of imaging systematic effects in the real data is more challenging than that of the linear-model based systematics in our mock tests. In this subsection we attempt to provide a quantitative evaluation of the residual systematics of DR7.

We quantify the residual systematics in the mean density against the 18 imaging maps using χ^2 of the mean density diagnostic as in Fig. 2.13 and Table 2.4 [13]. The null hypothesis is that the total residual squared error of the mean density observed in the data should be consistent with the distribution of the χ^2 statistics constructed with the null mocks. The

---

[13] To compare these with the mock results, we limit DR7 to the mock footprint, and therefore the χ^2 results are slightly different from Fig. 2.13 and Table 2.4. The mock footprint is smaller than the data footprint by almost a factor of two.

vertical lines in Fig. 2.23 present the χ^2 values observed in the data for before systematics treatment (9567.1) and after treatment with linear (2066.7), quadratic (1212.0), default Neural Network (767.3), and Neural Network plain (744.4) approaches. As a comparison, we present the distributions of the χ^2 observed in the null mocks (solid), contaminated mocks (dashed), and contaminated mocks after neural network mitigation (dot-dashed). Fig. 2.23 illustrates a factor of 12 improvement in terms of the residual χ^2 when using the neural network. Compared to the conventional quadratic method ($\chi^2 = 1212$), the NN-based method makes a factor of 1.6 improvement.

Figure 2.23: *Left*: χ^2 distribution for the null mocks (solid black), contaminated mocks (dashed black), and contaminated mocks with NN mitigation (dot-dashed black). The vertical dotted lines overlay the χ^2 values for the data with the default NN (red), quadratic (purple), and linear (grey) treatments. The χ^2 statistics before treatment is shown on the right (dark blue).

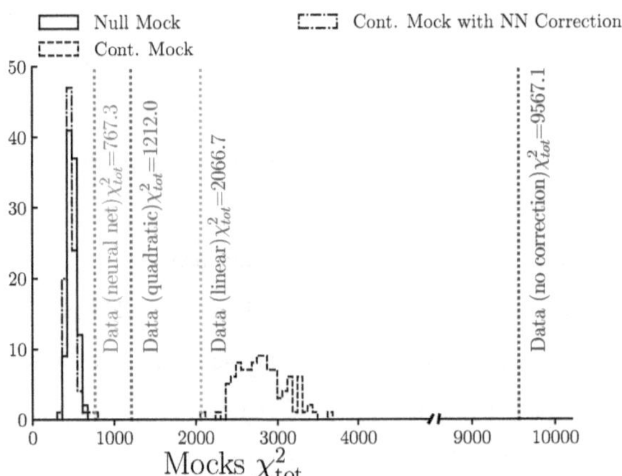

The mean and standard deviation for the distribution of χ^2 values observed in the null mocks are 487.1 ± 4.92, respectively. The same quantities observed in the contaminated mocks are 2819.0 ± 29.30, while Neural Network mitigation decreases these statistics to the mean of 472.0 ± 6.59. We perform Welch's t-test (Welch, 1947) on the χ^2 distributions of the null mocks and the mitigated contaminated mocks, and conclude that the two distributions have identical means with $t - statistic = 1.8$, $p - value = 0.06$ and the neural network recovers the underlying cosmological mean target density. Note that the mean χ^2 observed in the contaminated mocks, 2819.0, is much smaller than the value observed in the data (the blue vertical line, 9567.1). This indicates that the systematic effects rigorously simulated in the mocks are still not as strong as the systematics in the real data or the effect of the non-diagonal terms, that we ignored in the covariance, is different in the data and in the mocks.

We perform a hypothesis testing given the χ^2 value observed in the data after the neural network mitigation and the distribution of χ^2 values observed in the 100 null mocks. If the imaging maps were independent and normal deviates, the χ^2 values must follow a Chi-squared distribution for $dof = 360$ where 360 is the total number of bins. This assumption is not necessarily satisfied given the correlation among imaging systematics, and we indeed find χ^2 of the null mocks is better fit with $dof = 487$. Therefore, we assume that the underlying dof for evaluating χ^2 is 487. For DR7, the χ^2 decreased from 9567 before contamination to 767.3 after our NN-based mitigation. Compared to $dof = 487$ we expect for the approximate truth, this is a substantial excess, being very unlikely due to random fluctuation: for example, the top 5% of distribution with $dof = 487$ is at $\chi^2 = 539.4$, which is 40% lower than 767.3. Assuming a mere random fluctuation is bounded at this upper 5% and the χ^2 values of the data are consistent with that of the mocks, this offset could imply that we underestimated the covariance at least by $\sqrt{(767.3/539.4)} = 19\%$ or that our model, when applied to DR7, missed at least 19% of the systematic effects in the target density.

We further investigate the contribution of each imaging map to the residual systematics in DR7. Fig. 2.24 presents the χ^2 vs each imaging map after the linear (grey), quadratic (purple), neural network with feature selection (red), and neural network plain (blue) treatments. The statistics before mitigation are not shown for clarity. We also plot the 50- and 95-th percentiles of the same quantity observed in the null mocks in orange horizontal curves. Depth-g seems to be the main source of the residual systematics which is not mitigated even by the neural network-based methods. All methods also show residual systematics against skymag-g, skymag-z, and MJD-z. The standard treatments show some additional residual systematics against E(B-V), lnHI, and exptime-z. We perform further tests by masking out high extinction and low depth-g regions. We find that applying a more rigorous cut on depth-g (e.g., depth-g > 24.95) yields a more cleaner mean density at the expense of losing 9% of the data. Although our objective was to avoid applying rigorous cuts on imaging and demonstrate the gain by using non-linear methods, our tests suggest that careful masking based on imaging, particularly depth-g, seems necessary for cosmological studies.

In summary, our results indicate that any cosmological study requires a thorough analysis of systematics to determine an estimate of the residual systematic uncertainty for the parameters of interest. In order to use this sample for a cosmological exploitation, we would need to account for 19% (or more, depending on the assumption on the baseline) additional systematic errors to the statistical errors in the density field level. Our analysis also suggests that additional masking, especially based on depth-g, or improving the method to deal with depth and sky background issues would be the next steps to prepare this data for cosmological analysis. We leave this for future study, as the focus of this study is to compare our non-linear, neural network method to other map-based methods.

Figure 2.24: The breakdown of χ^2 values observed in the data on the mock footprint after linear (grey), quadratic (purple), neural network with feature selection (red), and neural network plain (blue) treatments. We also plot the 50- and 95-th percentiles of the same quantity observed in the null mocks in orange curves (note that $\chi^2(\mathbf{s})$ is not a continuous quantity). Given the 5% threshold, we can argue that there exists known residual systematics against E(B-V), depth-g, skymag-g, skymag-z, and MJD-z.

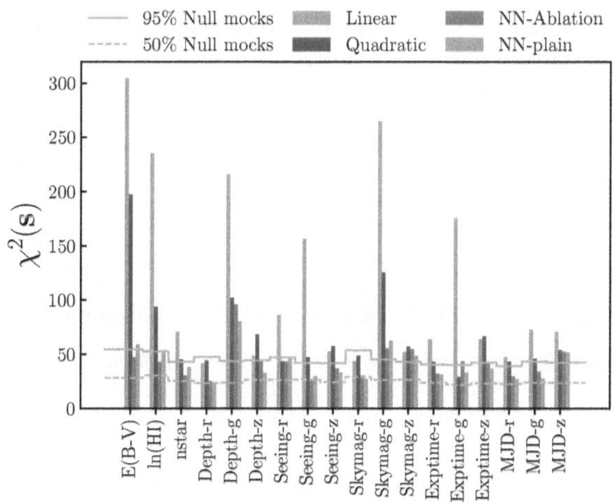

2.5 Conclusion

In this chapter, we have presented a rigorous application of an artificial neural network methodology to the mitigation of the observational systematics in galaxy clustering measurements of an eBOSS-like ELG sample selected from DR7 (see § 2.2). We have investigated the galaxy density dependency on 18 imaging attributes of the data (see Fig. 2.1). We compare the performance of the neural network with that of the traditional, linear and quadratic multivariate regression methods. The key aspects of our neural network methodology are:

- The application of k-fold cross-validation, which implements the training-validation-test split to tune the hyper parameters by evaluating how well the trained network generalizes to the unseen, validation data set and therefore to suppress overfitting when applied to the test set;

- The repeated split process until we cover the entire data footprint as test sets;

- The elimination of redundant imaging maps by the feature selection procedure to further reduce the overfitting problem and therefore protect the cosmological clustering signal.

We apply the output of our pipeline, i.e., the selection mask for the DR7 footprint to the observed galaxy density field. Benchmark selection masks are also produced employing the linear and quadratic polynomial regression. Comparing statistical results before and after applying the selection masks, we find that:

- Galactic foregrounds are the most dominant source of contamination in this imaging dataset (see Figs. 2.13, 2.14, and 2.15).

- This contamination causes an excess clustering signal in the auto power spectrum and correlation function of the galaxy density field on large scales (see Fig. 2.16).

- All mitigation techniques e.g., the neural network method as well as the linear multivariate models using the linear and quadratic polynomial functions, are able to reduce the auto and cross clustering signals (see Figs. 2.15 and 2.14);

- However, the neural network removes the excess clustering more effectively in the auto power spectrum and correlation function of galaxies (see Fig. 2.16).

The last result implies that our neural network method has a higher flexibility than both linear multivariate models we tested, and it is therefore capable of capturing the non-linear systematic effects in the observed galaxy density field.

We apply our methodology on two sets of 100 log-normal mock datasets with ('contaminated mocks') and without ('null mocks') imaging contamination to evaluate how well the ground truth cosmological clustering can be reconstructed in both cases, and therefore to validate the systematic mitigation techniques. All mitigation techniques are applied in the same way we treat the real data. The key results of our mock test are as follows:

- The feature selection procedure is able to identify most of the ten contamination input maps as important for the contaminated mocks while correctly identifying most of the maps as redundant for the null mocks (see Fig. 2.17).

- All three mitigation methods, i.e., the linear polynomial, quadratic polynomial, and neural network methods, perform similarly in terms of the residual bias in the presence of contamination. This is expected since the contamination model is based on the linear polynomial model which all three methods are capable of reproducing. The default neural network tends to slightly under-correct which is the outcome of the feature selection procedure. On the other hand, the linear and quadratic polynomial methods tend to slightly over-correct (see the right panel of Fig. 2.19).;

- In the absence of contamination, the neural network is the most robust against regressing out the cosmological clustering. This is mainly due to the feature selection process that appropriately reduces the flexibility of the mitigation (see the left panel of Fig. 2.19). Based on this result, we implement the feature selection procedure for DR7.

- Using χ^2 statistics, we quantify the bias and find that for the null mocks, the default neural network recovers the underlying clustering within 1σ C.L. (see Eq. 2.22) while the other methods return more than 4σ C.L. bias. For the contaminated mocks, all of the methods return biased clustering with $2.8-3.9\sigma$ C.L. which indicates that it is crucial for cosmological parameter estimation to determine the residual systematic uncertainty in scales sensitive to the parameters of interest (see the middle panels of Figs 2.19 and 2.22).

- All methods do not increase fractional variance during the mitigation process (see the bottom row of Fig. 2.19).

We also employ the mocks to investigate the remaining systematic effects in the data (Figs 2.13-2.15). While the neural network methods outperform the conventional methods, we conclude that the data exhibit around 19% residual systematics in the target number density (Figs 2.23 & 2.24). Our analysis suggests that a more rigorous masking on $depth-g$ (e.g., $depth-g > 24.95$) improves the mean density at the cost of losing 9% of data. To use this sample for cosmology, we therefore suggest a) accounting for 19% (or more, depending on the assumption on the baseline) additional systematic errors to the statistical errors in the density field level; b) performing further analysis of systematics and improvement of the mitigation method to deal with the depth and sky background issues.

To conclude, our analyses illustrate that the neural network method we developed in this chapter is a promising tool for the mitigation of the large-scale spurious clustering that is likely raised by the imaging systematics. Our method is more robust against regressing out the cosmological clustering than the traditional, linear multivariate regression methods. Such improvement will be particularly crucial for an accurate measurement of non-Gaussianity from the large-scale clustering of current eBOSS and upcoming DESI and the LSST surveys. Our method is computationally less intensive than other approaches such as the Monte Carlo injection of fake galaxies: analyzing DR7 using our default neural network method requires less than six CPU hours. Applications of our methodology on any imaging dataset would be straightforward. Our systematics mitigation methodology pipeline is publicly available at https://github.com/mehdirezaie/SYSNet.

3 Large-Scale Clustering of Quasars

We investigate the large-scale clustering of the final spectroscopic sample of quasars from the recently completed extended Baryon Oscillation Spectroscopic Survey (eBOSS). The sample contains 343708 objects in the redshift range $0.8 < z < 2.2$ and 72667 objects with redshifts $2.2 < z < 3.5$, covering an effective area of 4699 deg^2. We develop a neural network-based approach to mitigate spurious fluctuations in the density field caused by spatial variations in the quality of the imaging data used to select targets for follow-up spectroscopy. Simulations are used with the same angular and radial distributions as the real data to estimate covariance matrices, perform error analyses, and assess residual systematic uncertainties. We measure the mean density contrast and cross-correlations of the eBOSS quasars against maps of potential sources of imaging systematics to address algorithm effectiveness, finding that the neural network-based approach outperforms standard linear regression. Stellar density is one of the most important sources of spurious fluctuations, and a new template constructed using data from the Gaia spacecraft provides the best match to the observed quasar clustering. The end-product from this work is a new value-added quasar catalogue with the improved weights to correct for nonlinear imaging systematic effects, which will be made public. Our quasar catalogue is used to measure the local-type primordial non-Gaussianity in our companion paper, Mueller et al. in preparation.

3.1 Introduction

Quasar clustering measurements at large scales are sensitive to spurious density fluctuations caused by imaging properties (Ross et al., 2013; Ho et al., 2015; Ross et al., 2012b; Pullen and Hirata, 2013; Leistedt et al., 2013; Leistedt and Peiris, 2014; Kalus et al., 2019; Laurent et al., 2017), and are the primary source of systematic error that impede a robust analysis of primordial non-Gaussianity. The reasons for data having systematic errors include, but are not limited to, Galactic foregrounds (such as Galactic extinction and stellar contaminations), seeing, and survey depth variations. These properties fluctuate across the survey footprint and introduce spurious variations in the observed density field of quasars. Because many of these effects have large-scale variations, they result in an excess clustering signal at large scales. For instance, Pullen and Hirata (2013) analyzed the SDSS DR6 quasar sample (Richards et al., 2008) to find that systematic error in the sample does not allow a robust inference on $f_{\rm NL}$. They found the 2% RMS fluctuations in the quasar density, while they argued that the fluctuations under 1%(0.6%) is required to measure $f_{\rm NL}$ less than 100 (10).

Conventional techniques for improving quasar clustering measurements rely on a regression analysis of observed quasar density against a set of mappable properties that describe imaging conditions during observing (e.g., Ross et al., 2011; Ross et al., 2012b; Delubac et al., 2016; Prakash et al., 2016; Ross et al., 2017b; Laurent et al., 2017; Bautista et al., 2018). This approach has been further enhanced to account for inter-correlations among imaging variables, validated with simulations, and applied to other photometric surveys, such as the Dark Energy Survey (e.g., Elvin-Poole et al., 2018b; Wagoner et al., 2020). The regression analysis fits for spurious fluctuations in target quasar density and then is employed as a selection function to assign an appropriate weight to each target to eliminate such variations. The method of mode-projection removes the modes that strongly correlate with imaging templates in covariance based estimators (e.g., Tegmark, 1997; Leistedt et al., 2013; Leistedt and Peiris, 2014; Kalus et al., 2019). Regression and mode-projection turn out to be mathematically equivalent (Kalus et al., 2016). An alternative technique uses cross-correlations of multiple tracers, or same tracer at different redshift bins, with the assumption that each sample responds differently to imaging systematics and thus cross-correlations are not affected (e.g., Rhodes et al., 2013). For a benchmark analysis of various cleaning methods, see, e.g., Weaverdyck and Huterer (2020).

The extended Baryon Oscillation Spectroscopic Survey (eBOSS; Dawson et al., 2016) is part of the fourth phase of SDSS. The eBOSS program is the final large-galaxy redshift survey proposed within SDSS aiming to measure the expansion history and energy contents of the Universe using Large-Scale Structure (LSS) (Ahumada et al., 2020). With a series of galaxy redshift surveys, SDSS has probed the distribution of matter traced by galaxies and quasars using the dedicated 2.5m Sloan Foundation Telescope (Gunn et al., 2006) at the Apache Point Observatory from 1998 to 2019. During four distinct phases, SDSS collected images and spectra of thousands to millions of astronomical objects and created the largest three-dimensional map of the cosmic web to date. The first phase of SDSS led to the detection of Baryon Acoustic Oscillations (BAO) in the large-scale clustering of galaxies (Eisenstein et al., 2005), at the same time as the 2-degree Field Galaxy Redshift Survey (Cole et al., 2005). Calibrated by CMB measurements, the BAO scale has been widely utilized as a standard ruler leading to robust distance-redshift measurements and constraints on the nature of Dark Energy with LSS surveys. The observations of the BAO signal in LSS have improved since the early detections, passing the 5σ detection threshold and now cover a wide range in redshift using only SDSS data (e.g., Percival et al., 2010; Anderson et al., 2014a; Alam et al., 2017).

In the fourth phase of SDSS, eBOSS adopted the same 1000-fiber spectrograph (Smee et al., 2013) from its predecessor, BOSS (Dawson et al., 2012), and utilized improved pipelines for redshift estimation, background subtraction, and flux calibration (Bolton et al., 2012; Hutchinson et al., 2016; Jensen et al., 2016; Bautista et al., 2017). eBOSS introduced a new class of targets for actively star-forming galaxies with strong [OII] emission lines, known as Emission-Line Galaxies (ELGs; Raichoor et al., 2017). ELGs extend over the redshift range of $0.6 < z < 1.1$ (Raichoor et al., 2020) and fill the redshift gap between Luminous Red Galaxies (LRGs; Prakash et al., 2016) and Quasi-Stellar Objects (QSOs; Myers et al., 2015), referred to as quasars in this Chapter.

We present a careful assessment and treatment of imaging systematic effects in the final sample of quasars (Lyke et al., 2020; Ross et al., 2020) from the eBOSS Data Release 16 (Ahumada et al., 2020), the largest sample of quasars available to date. We improve upon a neural network-based cleaning approach, which was originally developed, validated, and applied to eBOSS-like emission-line galaxies in Rezaie et al. (2020). Compared to emission-line galaxies, quasars represent a class of sparser targets for galaxy surveys, particularly for spectroscopic surveys. This Chapter enhances the method to deal with the sparsity of the eBOSS DR16 quasars by modeling quasar counts per pixel with the Poisson distribution. We improve neural network training to be less prone to local minima by utilizing a cyclic learning rate. We perform a comprehensive benchmark of linear and nonlinear treatments, and investigate each method effectiveness in reducing spurious fluctuations. Residual systematic errors are quantified using mean quasar density contrasts and cross-power spectra against imaging templates. For the significance of residual spurious fluctuations, we construct covariance matrices from realistic simulations. Our primary objective is to examine, quantify, and mitigate the potential sources of observational systematic error and enhance quasar power spectrum measurements for constraining primordial non-Gaussianity. This work presents a new set of value-added catalogues with the enhanced systematic weights to account for imaging systematic effects more exquisitely compared to the standard linear regression which is used in the eBOSS pipeline. The new quasar catalogue is utilized in two accompanying papers for constraining the local-type primordial non-Gaussianity (Mueller and et al. in prep., 2020) and for exploring the impact of imaging systematic error on Baryon Acoustic Oscillations (Merz et al., 2021).

This Chapter is structured as follows. Section 3.2 describes the eBOSS quasar sample and simulated datasets used in this work. Section 3.3 outlines the power spectrum estimator and our strategies for the treatment and characterization of imaging systematics.

In Section 3.4, we present our statistical tests for quantifying residual systematic errors, assess the performance of different cleaning methods, and illustrate the impact of imaging properties on the measured power spectrum of the DR16 sample and that of the simulated catalogues. Finally, we conclude with a summary of our results and their significance for constraining primordial non-Gaussianity with quasar clustering in Section 3.5.

3.2   Data

This section describes the final sample of quasars from the completed eBOSS DR16 dataset (Ross et al., 2020; Lyke et al., 2020) and the simulated EZmock catalogues (Zhao et al., 2021) used in our analysis.

3.2.1   eBOSS DR16 Quasars

The DR16 quasar sample is the final release of large-scale structure data from SDSS-IV eBOSS and provides twice the number of quasars and sky coverage over the previous Data Release DR14. The target selection of quasars for eBOSS is presented in Myers et al. (2015), and only the main details are briefly summarized here. It uses optical and infrared imaging, respectively, from SDSS and WISE (Wide-Field Infrared Survey Explorer; Wright et al., 2010). The photometric data are taken in five bands (u, g, r, i, z; Fukugita et al., 1996) and calibrated to account for the Galactic dust effect using correction factors presented in Schlafly et al. (2012). The ability to obtain a quasar with redshift $z > 0.9$ is improved by 20% using the xDQSOz algorithm (Bovy et al., 2012). The magnitude selection applies the extinction-corrected flux cuts in the g and r bands, namely $g < 22$ and $r < 22$, to choose the CORE eBOSS quasars. The targeting strategy enables the observation of the Ly-α high-z quasars by relaxing the maximum redshift cut and reduces stellar contaminations in the sample by incorporating a mid-infrared cut. The redshifts are estimated using the

REDVSBLUE[14] principal component analysis algorithm described in Lyke et al. (2020), with 95% completeness and 2% false positive rate. The main quasar sample spans the redshift range of $0.8 < z < 2.2$ and is used for various cosmological analyses of the BAO and RSD features (Hou et al., 2021; Neveux et al., 2020; du Mas des Bourboux et al., 2020; Smith et al., 2020). Additionally, the high-z quasars with $2.2 < z < 3.5$ are employed for measuring the BAO signal in the Ly-α forest (Chabanier et al., 2019; Blomqvist et al., 2019; de Sainte Agathe et al., 2019).

The preparation of the large-scale clustering catalogue of quasars is outlined in Ross et al. (2020). As well as associating attributes and *weights* per object, the process includes generating a set of unclustered synthetic objects (often referred to as the *random catalogue* or *randoms*) matching the expected weighted density of quasars and accounting for the radial and angular survey geometry. Standard procedures account for the veto masking, completeness cuts, fiber collision correction by nearest neighbour upweighting, redshift failure correction through weighting, and imaging systematics by linear regression against templates. The products are the tabulated coordinates of the quasars and randoms along with appropriate columns for per-object weights, separately for the North Galactic Cap (NGC) and South Galactic Cap (SGC) regions. Each weight column is intended to address a particular systematic effect in the observed data, which is discussed briefly in the following.

The physical size of a fiber limits the ability to observe a pair of quasars within 62 *arcsec* of separation. This *fiber collision* effect is not random and affects quasar clustering. Up-weighting the nearest neighbor corrects for much of the large-scale effect, although small-scales remain affected and require a more complicated procedure to achieve unbiased removal (Hahn et al., 2017; Bianchi and Percival, 2017; Mohammad et al., 2020). In our analysis we use the simple nearest neighbour upweighting scheme as we are only interested in large scales (as used, for example, by Neveux et al., 2020).

---

[14]https://github.com/londumas/redvsblue

The completeness of redshift estimation varies among the fibers across the spectrographs, with fibers near the edge having a higher rate of redshift measurement failures. The weight w_{noz} is determined to mitigate this systematic error by weighting by the reciprocal of the probability that each fiber accurately measures a redshift. The impact of redshift completeness on quasar clustering is investigated in Hou et al. (2021).

The remaining source of systematic uncertainty is associated with the properties of the imaging data from which targets were selected including, but not limited to, stellar contamination, Galactic extinction, and inaccurate photometric calibrations. The standard method uses a multivariate linear model to find the intercorrelation between target density and imaging templates, and provides a per-object weight w_{systot} to reduce this systematic effect (Ross et al., 2012b; Bautista et al., 2018). We explain the standard treatment and how the systematic weight w_{systot} is obtained in §3.3.1. Collectively, to account for all of these observational effects, each quasar and random object must be weighted by,

$$w = w_{\mathrm{systot}} \times w_{\mathrm{noz}} \times w_{\mathrm{FKP}} \times w_{\mathrm{cp}}, \tag{3.1}$$

where $w_{\mathrm{FKP}} = [1 + n(z)P_0]^{-1}$ is the FKP weight (Feldman et al., 1994) with $P_0 = 6000\ \mathrm{Mpc}^3 h^{-3}$ based on the expected power[15] on scales around $0.01 - 0.3\ h/\mathrm{Mpc}$, and $n(z)$ is the number density at redshift z. We require completeness > 0.5 (for both parameters COMB_BOSS and SECTOR_SSR[16]) as well as $0.8 < z < 3.5$ to avoid low-quality samples. The quasar catalogue contains 343708 objects in the redshift range $0.8 < z < 2.2$ and 72667 in the redshift interval $2.2 < z < 3.5$, covering an effective area of 4699 square degrees.

---

[15]This is the default value for the BAO and RSD studies with eBOSS (e.g., Hou et al., 2021; Neveux et al., 2020). The impact of the FKP weight on f_{NL} constraints is investigated in our companion paper, Mueller and et al. in prep. (2020).

[16]These parameters describes the survey completeness and the probability of detecting an object with a good redshift in spite of other instrumental factors.

Hereafter, we refer to the sample with $0.8 < z < 2.2$ as *main*, and the one with $2.2 < z < 3.5$ is referred to as *high-z*.

Figure 3.1: Mean density of the DR16 quasars as a function of redshift for the NGC (solid blue) and SGC (dashed orange) regions. The vertical dotted lines represent the redshift cuts for selecting the main and high-z samples.

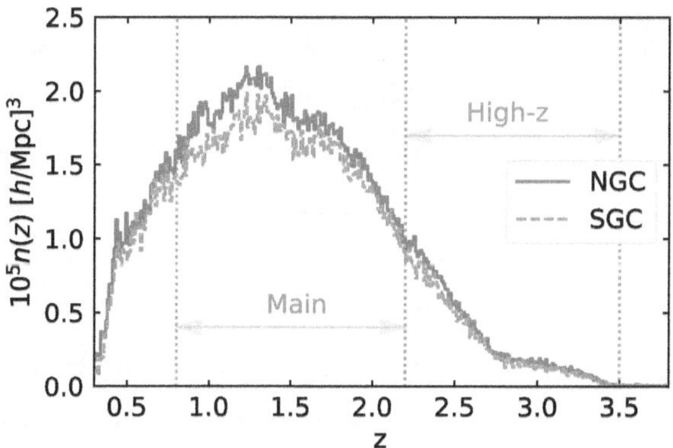

Fig. 3.1 shows the mean number density of quasars from the DR16 catalogue as a function of redshift for the NGC (solid blue) and SGC (dashed orange) regions. The main and high-z samples are separated by vertical dotted lines. Targeting efficiency varies slightly between the two caps which causes small differences in $n(z)$. Figure 3.2 shows the sky coverage of the DR16 quasars in equatorial coordinates. The top panel is the Mollweide projection[17] of the quasar density field in deg^{-2} after the standard treatment. The solid red curve represents the Milky Way plane. The effective area of the NGC and SGC regions are 2860 and 1839 deg^2, respectively (see table 3 in Ross et al., 2020). The quasar density in

---

[17]The density map is in HEALPIX (Gorski et al., 2005) with NSIDE = 128, which corresponds to a pixel area of 0.21 deg^2

each pixel is corrected for pixel completeness using the number of random objects. The bottom panel describes the residual quasar density between the standard and new neural network treatments. These methods are thoroughly described in §3.3.1. Interestingly, the residual map indicates that the size of the correction depends on angular direction, with the nonlinear approach leading to more correction in particular regions.

3.2.2 Synthetic Catalogues

We use synthetic catalogues (Zhao et al., 2021), generated using the extended Zeldovich approximation (Zel'Dovich, 1970; Chuang et al., 2015), to construct covariance matrices of our statistics and perform robustness tests, characterizing the significance of residual systematic uncertainties. Throughout this manuscript, we refer to these mocks as the *EZmocks*. The EZmocks are tuned to reproduce accurate two-point clustering statistics, e.g., within 1% of an N-body simulation on $k < 0.55\ h/\mathrm{Mpc}$ and $r > 10\ \mathrm{Mpc}/h$ (Chuang et al., 2015), and thus are suitable for studying the large-scale clustering of galaxies and quasars. These simulations are sufficient for this work since the impact of imaging systematics on quasar power spectrum measurements appears primarily on long-wavelength modes, e.g., $k < 0.01\ h/\mathrm{Mpc}$. A detailed description for the creation of the eBOSS EZmocks is presented in Zhao et al. (2021).

An EZmock realization is created by adopting the Zeldovich approximation to generate a density field. Then, stochastic and parametric techniques are applied to simulate the scale-dependent nonlinear biasing effects. A flat ΛCDM universe defined by cosmological parameters $h = 0.678$, $\Omega_M = 0.307$, $\Omega_\Lambda = 0.693$, $\Omega_b = 0.0482$, $\sigma_8 = 0.822$, and $n_s = 0.961$ is considered as the input cosmology for the EZmocks. The mock realizations do not contain primordial non-Gaussianity, i.e., $f_{\mathrm{NL}} = 0$. Each mock catalogue is constructed by combining seven periodic boxes with the comoving side length $L = 5\ \mathrm{Gpc/h}$ to mimic a light-cone geometry which accounts for the redshift evolution of

Figure 3.2: *Top*: Mollweide projection of the DR16 sample in the redshift interval of $0.8 < z < 3.5$ and the effective area of 4700 deg^2 with the standard treatment for imaging data systematics. The solid red curve represent the Galactic plane. *Bottom*: The difference in corrected quasar density between applying the standard or neural network treatments for imaging data systematics.

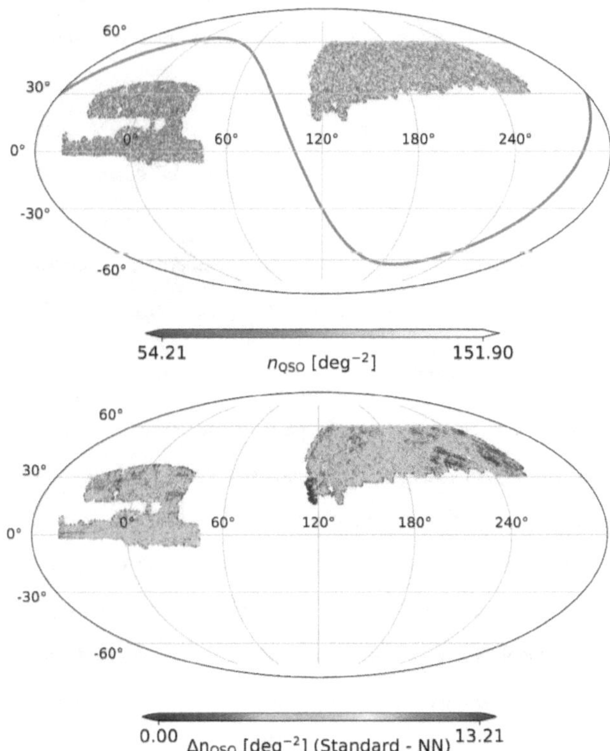

quasars. Then, the mock catalogue is sub-sampled along the line of sight to simulate the redshift distributions of the DR16 quasar sample.

We use two sets of simulations in this Chapter. Each set contains roughly 1000 independent realizations per Galactic cap. We have the mocks only for the redshift interval $0.8 < z < 2.2$. One set of the mocks is manipulated to reflect the effects of observational systematics, including fiber collision, stellar contamination, redshift failures, and angular imaging systematics. While the other set lacks any observational systematics, except the survey geometry effect. Throughout this manuscript, we refer to these two sets as the *contaminated* or *null* mocks, indicating whether or not observational systematic effects are added to the simulations. In particular importance to our work, angular imaging systematics are simulated based on a linear multivariate fit to the DR16 quasar density against two imaging templates[18] for the SDSS stellar density and depth-g with HEALPIX resolution of NSIDE=512 (see 3.2.3). Similar to the DR16 sample, eBOSS pipelines are applied to the EZmock realizations to calculate and assign the appropriate attributes, which are intended to mitigate the simulated systematic effects and recover the ground truth clustering.

3.2.3 *Imaging Templates*

In total we have 17 templates of photometric variables, that are involved in the SDSS imaging and targeting, as potential sources of systematic uncertainties. These templates comprise three maps tracking the structure of the Milky Way: *Galactic extinction* (Schlegel et al., 1998), *neutral hydrogen column density* (HI4PI Collaboration et al., 2016), and *stellar density* from either the SDSS program (as used in Bautista et al., 2018) or the Gaia spacecraft (Gaia Collaboration et al., 2018). The survey-specific templates are *seeing*, *sky brightness*, and *survey depth* in four bands (*griz*). The remaining two maps are *run* and *airmass*.

---

[18]This is a conservative approach for simulating imaging systematics in the mock catalogues since the treatment of the real sample has shown that more than two templates are required to achieve a desirable cleaning.

We produce these imaging templates in HEALPIX by using a catalogue of uniformly distributed random objects painted with imaging features. For each pixel, the average of each imaging property is computed over the random objects in the pixel. The maps are created in the ring ordering format and two resolutions, NSIDE=256 and 512, respecively corresponding to 0.23 and 0.11 degrees. Although the standard treatment uses templates with NSIDE=512 for cleaning the eBOSS DR16 quasars, the preparation of templates in a different resolution, e.g., 256, will enable us to investigate how changing pixel size might influence the mitigation process and quasar clustering measurements.

A caveat of all template-based cleaning methods is that the available templates are not perfect, and this limitation might affect the performance of such treatments; for instance, the default stellar density map is an incomplete sample of stars, which is constructed as the number of type PSF objects with $i < 19.9$ from SDSS. Therefore, our proxy of stellar artifacts may be an incomplete representation of existing stellar contaminations in the DR16 sample. Indeed, as we discuss later in §3.4, we find a noticeable reduction in residual systematic error by swapping the SDSS map for a different stellar map from the Gaia DR2 with $12 < g < 17$ (Gaia Collaboration et al., 2018). The other caveat is that using all imaging maps for training might add the input noise to the neural network, increase chance cross-correlation between cosmological signal and imaging fluctuations, and reduce the effective number of modes. Therefore, we decide to follow a conservative approach to use a minimum number of templates in regression analysis. Later in §3.4, our results show that spurious fluctuations against imaging properties are alleviated by using only fives maps including *sky-i*, *seeing-i*, *extinction*, *depth-g*, and *Gaia stellar density*. Based on the χ^2 of the mean density residuals, the first four maps are identified as the primary sources of systematic error in the standard cleaning procedure (Ross et al., 2020), and hereafter we refer to this set as the *known* templates.

3.3 Analysis Techniques

This section presents the techniques and statistics employed in this work for the characterization and mitigation of residual spurious fluctuations, and outlines the estimator for measuring the large-scale clustering of the DR16 sample and that of the EZmock realizations.

3.3.1 Template-based Mitigation

3.3.1.1 Linear Multivariate Regression

The default systematic weights in the DR16 catalogue (Ross et al., 2020) are obtained based on a linear multivariate regression using the four known maps, i.e., *sky-i*, *seeing-i*, *extinction*, and *depth-g*, as the primary sources of systematic fluctuations. Linear regression is the most commonly used approach to reduce the effects of imaging systematics in previous SDSS catalogues (e.g., Ross et al., 2011; Ross et al., 2012b; Myers et al., 2015; Prakash et al., 2016; Ross et al., 2017b). Despite some minor differences among the various implementations, a common assumption is that the observed number density of quasars in pixel i is a linear combination of imaging quantities in the pixel (Bautista et al., 2018, eq. 6),

$$y_i = \bar{n} + \sum_{j=1} p_j x_{j,i}, \tag{3.2}$$

where $\bar{n}$ is the mean density of quasars, $x_{j,i}$ is the j'th imaging variable in pixel i and p_j is the corresponding linear coefficient. These methods mostly differ either on their cost function[19] or on the number of templates used in the model. For instance, the least square error is evaluated over either the binned or pixelated mean density of quasars. Sometimes, the regression routine fits against all templates simultaneously or one template at a time in each iteration.

---

[19]Cost function is used to find the best parameters of a model.

Specifically, the method presented in Bautista et al. (2018) implements a simultaneous fit using the four known maps as the independent variables in the model and chooses the least square error on the binned quasar density as the cost function. A subtle aspect of the cost function is that the residual fluctuations are summed over not only the four known maps but also the SDSS stellar density and airmass. This design will train the model parameters to explain the trends against the six templates while providing only the four known maps as input. Given the four known maps as the input variable x, the systematic weight of quasars in pixel i is defined as, $w_{\mathrm{systot},i} = 1/(c + \sum_j p_j x_{j,i})$, where c is the intercept. The optimal coefficients and intercept c are then derived by minimizing the least-squares error[20] on the binned quasar density after applying the systematic weight w_{systot} to each quasar object. The binned density is accounted for other observational and pixel completeness effects by weighting the quasars and randoms respectively by $w_{\mathrm{tot,g}}$ and $w_{\mathrm{tot,r}}$, see, Eq. 3.7.

3.3.1.2 Non-Linear Approximation using Neural Networks

The DR16 quasar sample is very sparse, with a surface density of 73 deg^{-2} in the redshift range $0.8 < z < 2.2$ (main) and 15 deg^{-2} in $2.2 < z < 3.5$ (high-z). The high sparsity poses a major challenge for regression-based treatments by making it difficult to disentangle the effect of systematics and the cosmological signal. This Chapter expands the neural network-based methodology described in Rezaie et al. (2020) to obtain nonlinear systematic weights. In what follows, we first describe the specifics of the neural network architecture, and then elaborate on how we further improve the methodology specifically with the implementation of the Poisson statistics in the cost function to properly deal with the high sparsity eBOSS QSO data and a cyclic learning rate to enhance training against local minima.

---

[20]It is worth noting that unlike the neural network method, this approach uses all data for training and does not apply any form of training, validation, and testing. We believe linear regression is not in a limit prone to over-fitting, i.e., the degree of freedom is much smaller than the number of data points.

Feed forward neural networks: Neural networks are universal approximators that can model a wide variety of non-linear mappings from a set of independent variables to a target variable. Neural networks can account for nonlinear spurious fluctuations caused by imaging and the cross-correlation among imaging properties. We use y_i and $\mathbf{x}_i$ to respectively denote the observed quasar count and imaging features[21] in pixel i. We implement a fully connected feed-forward neural network to model the observed quasar count y_i from imaging templates as input features $\mathbf{x}_i$. At its standard configuration, a fully connected feed-forward neural network is a system of neurons organized in a series of interconnected layers, where each neuron is connected to all neurons in the previous layer and the following layer[22]. For instance, the output value of neuron μ in layer $l + 1$, a_μ^{l+1}, is obtained from the linear combination of the output values from the previous layer neurons, a_ν^l, after applying a nonlinear activation function f,

$$a_\mu^{l+1} = \psi(b_\mu^l + \sum_\nu w_{\mu\nu}^l a_\nu^l), \tag{3.3}$$

where $w_{\mu\nu}^l$ and b_μ^l are respectively the associated weights and bias connecting to neuron μ. The nonlinear function ψ determines how much of the signal travels from layer l to $l + 1$. In this notation, the first layer input values are imaging properties x_i and the last layer output is then compared to y_i. The network architecture comprises three fully-connected hidden layers[23] with twenty neurons with the rectified linear activation function, $\psi(u) = \max(0, u)$, on each hidden layer and a single neuron with the softplus activation function, $\text{Softplus}(u) = \log[1 + \exp(u)]$, on the output layer. The softplus activation in the last layer will ensure the network output is always positive. We also add a batch

---

[21] We use the z-score normalization scheme to standardize our imaging templates, i.e., $z = (x - \mu)/\sigma$ where μ and σ are the mean and standard deviation of imaging feature x determined from the training set.

[22] Bias neurons are an exception and connect only to the subsequent layer.

[23] We follow a grid search approach to experiment with a various number of hidden layers ranging from two up to five hidden layers, however we do not observe any significant change in validation loss. Therefore, we fix the architecture at three hidden layers with 20 units on each hidden layer.

normalization layer after each layer, except the output layer. Batch normalization stabilizes and expedites training by scaling the inputs to each layer to have a mean of zero and a standard deviation of one (e.g., Ioffe and Szegedy, 2015).

Poisson statistics in the cost function: For the DR16 quasars, we decide to use the negative Poisson log-likelihood as the default cost function and the Mean Squared Error for benchmarking. Assuming the quasar counts per pixel are independent and identically distributed variables, we can write the joint probability for N pixels as,

$$L = f(y_1, ..., y_N|\theta) = \prod_{i=1}^{N} f(y_i|\theta, \mathbf{x}_i), \tag{3.4}$$

which is a reasonable assumption provided that the input features $\mathbf{x}_i$ do not include any spatial information, and we do not intend to provide any feature that contains cosmological clustering in our modeling. Otherwise, the mitigation procedure will learn the clustering signal and remove it as well. The objective is then to find the best set of parameters θ that maximizes the likelihood L, or minimizes its negative logarithm $-\log L$. Assuming y_i follows a Poisson distribution, we have

$$f(y_i|\theta, x_i) = \frac{\lambda(\theta, \mathbf{x}_i)^{y_i} \exp^{-\lambda(\theta, \mathbf{x}_i)}}{y_i!}, \tag{3.5}$$

where λ is the expected number of quasars, $\lambda > 0$ by definition, and a function of imaging features. We use $\lambda_i \equiv \lambda(\theta, \mathbf{x}_i)$ for brevity and obtain the negative log likelihood[24] as the cost function J,

$$J \equiv -\log(L) = \sum_{i=1}^{N}[\lambda_i - y_i \log(\lambda_i)]. \tag{3.6}$$

In this notation, the quasar number count y_i is assumed to be an integer value. This assumption breaks down in practice since each object in the DR16 catalogue needs to be weighted for other observational systematics and pixel completeness beforehand, and that turns y_i into a non-integer value. To circumvent this issue and still account for the pixel

---

[24] We omit the term $\log(y_i!)$ from our objective function since it does not depend on θ.

completeness and other observational effects, we decide to use the raw number count of quasars for y_i, and instead weight the model prediction in pixel i, i.e., λ_i, by the ratio of the weighted number of randoms to that of quasars in that pixel. To this end, the quasar and random objects are respectively weighted by $w_{\text{tot,g}}$ and $w_{\text{tot,r}}$,

$$w_{\text{tot,g}} = w_{\text{noz}} \times w_{\text{cp}} \times w_{\text{FKP}},$$

$$w_{\text{tot,r}} = w_{\text{FKP}} \times \text{comp}_{\text{BOSS}}, \tag{3.7}$$

where $\text{comp}_{\text{BOSS}}$ is the survey completeness that describes the likelihood of obtaining a good redshift, regardless of any other instrumental deficiencies.

Cyclic learning rate: The weights $w_{\mu\nu}^l$ and biases b_μ^l of the neural network are trained using the Decoupled Weight Decay Regularization optimizer (AdamW; Loshchilov and Hutter, 2017), which is an iterative gradient descent approach for optimization. Specifically, the parameters are updated in the opposite direction of the gradient of the cost function with respect to the parameters,

$$m_{t+1} = \beta_1 m_t + (1 - \beta_1)\nabla J(\theta_t), \tag{3.8}$$

$$v_{t+1} = \beta_2 v_t + (1 - \beta_2)[\nabla J(\theta_t)]^2, \tag{3.9}$$

$$\theta_{t+1} = \theta_t - \eta_t \frac{m_{t+1}}{\sqrt{v_{t+1}} + \epsilon}, \tag{3.10}$$

where $\epsilon = 10^{-8}$ and the learning rate η_t controls the magnitude of each parameter update per iteration t. The first and second moments of the gradients, i.e., m_t and v_t, are initialized as zero. The parameters β_1 and β_2 determine the average history of the first and second moments of the gradients, and are fixed at 0.9 and 0.999, respectively (see, e.g., Ruder, 2016, for a review of gradient descent methods).

We incorporate a cyclic learning rate to prevent the optimizer from being trapped in a local minimum. Specifically, the learning rate η at epoch t is scheduled to vary following the method presented in (Loshchilov and Hutter, 2016),

$$\eta_t = \eta_{\min} + \frac{1}{2}(\eta_{\max} - \eta_{\min})[\cos(\frac{T_{\text{cur}}}{T_i}\pi) + 1], \tag{3.11}$$

where $\eta_{\max}$ is the initial (maximum) learning rate, $\eta_{\min}$ is the minimum learning rate, T_{cur} is the number of epochs since the last restart, and T_i is the number of epochs between two subsequent restarts. The neural network training begins with $T_{i=0} = 10$, but then we increase T_i by a factor of two after each restart (i.e., T_i changes like 10, 20, 40, ...). We follow the procedure presented in Smith (2015) to search an exponential grid[25] for the optimal values of $\eta_{\min}$ and $\eta_{\max}$. The optimal values are chosen such that the variation of loss per training epoch is maximized. We separately perform the learning rate finding procedure for each Galactic cap, redshift selection, template resolution since the balance between systematics and noise changes.

Cross-validation: We apply five-fold cross-validation to perform training, validation, and testing the network on the entire footprint, while ensuring no overlap among any of the training, validation, or testing sets. Specifically, we randomly split the entire footprint into 60% training, 20% validation, and 20% testing sets. The training set is used to compute the gradients and update the parameters. In each epoch, the model is applied to the validation set to assess the prediction error when the model is applied on unseen data. We let the networks to train for 150 epochs, and finally we apply the model with the lowest validation error on the test set. By changing the arrangement of the training and validation sets, the neural network will be tested on the entire footprint. Similar to Ross et al. (2020), we perform training, validation, and testing for the main and high-z samples in the NGC and SGC regions separately and finally aggregate the results. This approach assumes that the effects of imaging systematics along the line-of-sight do not vary much. We relax this assumption later by further splitting the main sample into $0.8 < z < 1.5$ and $1.5 < z < 2.2$,

---

[25]github.com/davidtvs/pytorch-lr-finder; github.com/fastai/fastai

Figure 3.3: The Pearson correlation coefficient between the averaged predicted quasar counts from neural networks for various ensemble size in the NGC (black) and SGC (red) regions, and the main and high-z samples. The correlation increases beyond 90 per cent after averaging over three networks.

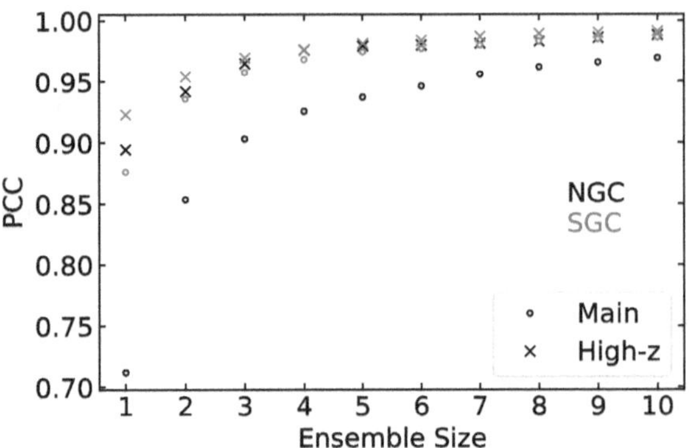

and analyzing each subsample separately. As shown later in §3.4, we find no significant difference. Therefore, we leave an extensive 3D regression of target density and imaging properties for future work.

Unlike linear regression, the objective function of a neural network is non-convex, and thus there is no unique set of best-fit parameters. This results in an intrinsic scatter on the predicted quasar density from a single neural network. We create an ensemble of 20 networks, through random initializations of the parameters, to mitigate the issue of model uncertainty, reduce the dispersion of the predicted quasar counts, and increase the stability of the selection function. We then take the mean predicted number of quasars across all networks in the ensemble as the final selection function. Fig. 3.3 illustrates the Pearson correlation coefficient (PCC[26]) between the predicted quasar counts of two independent ensemble subsets as a function of the ensemble size for the main and high-z samples in the NGC (black) and SGC (red) regions. The Pearson coefficient illustrates that the mean predicted densities between two ensembles, each contains only four or more networks, correlate more than 90 per cent. Then, the systematic weight for quasars in pixel i is defined by $w_{\mathrm{systot}} \propto 1/\lambda(\theta, \mathbf{x}_i)$, where w_{systot} is normalized such that the total weighted number of quasars stays the same as before treatment [27]. The figure shows that with our default ensemble size of 20, we are introducing very little intrinsic scatter in the process of the neural network.

3.3.2 Residual Systematics Tests

3.3.2.1 Mean Density Contrast

This test is sensitive to the uniformity of the DR16 sample by clustering pixels with similar imaging properties. We compute the mean density contrast of quasars against each imaging variable to quantify the deviations and fluctuations in the sample before and after cleaning. For a given imaging quantity x_j, the mean density contrast δ is,

$$\delta(x_j) = \frac{\sum_i n_{g,i}}{\alpha \sum_i n_{r,i}} - 1, \tag{3.12}$$

where $n_{g,i}$ and $n_{r,i}$ are the weighted number of quasars and randoms in pixel i; α is the factor to normalize the number of randoms to that of quasars; and the summation $\sum_i$ is computed over pixels with $x_{j,i} \in [x_j, x_j + \Delta x_j]$. We adopt an equal frequency binning scheme in which pixels are initially sorted from the minimum to maximum imaging value, and then

---

[26]$\mathrm{PCC}(X, Y) = cov(X, Y)/\sqrt{cov(X, X)cov(Y, Y)}$, where $cov(X, Y)$ is the covariance between variable X and variable Y.

[27]We also limit the systematic weights to $0.5 < w_{\mathrm{systot}} < 2.0$ to avoid extreme corrections. For pixels where we do not have imaging information, we use the mean value over nearest neighbors to estimate the systematic weight. We do not observe a significant impact on quasar clustering.

the width Δx_j is varied such that each bin contains the same area, i.e., the same number of randoms. By design, this statistic is not sensitive to the underlying true power spectrum, since the mean of density contrast is computed over a large number of pixels in the imaging phase space. Therefore, the cosmological component in δ cancels out.

The quasar and random objects are weighted differently before and after mitigation; initially, the quasars are weighted by $w_{\text{tot,g}}$ and the randoms are weighted by $w_{\text{tot,r}}$, Eq. 3.7. After mitigation, we use Eq. 3.1 to weight both the quasars and randoms. Finally, we compute the mean density contrast against all 17 templates, $\delta = [\delta(x_1), \delta(x_2), ..., \delta(x_{17})]$, and quantify the total mean density residuals with χ^2 statistics,

$$\chi^2_{\text{tot}} = \delta^\dagger C^{-1} \delta, \tag{3.13}$$

where the covariance matrix C is estimated from the null EZmock mean density contrasts, $C =< \delta^\dagger \delta >$, where the angle brackets represent ensemble average over the EZmock realizations. The estimated covariance matrix is then unbiased by the Hartlap factor (Hartlap et al., 2007),

$$C = \frac{N_{\text{mocks}} - 1}{N_{\text{mocks}} - N_{\text{bins}} - 2} C_{\text{biased}}, \tag{3.14}$$

where N_{mocks} is the number of mock realizations, e.g., 1000 for the NGC and 999 for the SGC, and N_{bins} is the number of imaging bins, i.e., 136. Fig. 3.4 shows the estimated covariance matrix of the mean density contrast from the 1000 null EZmock realizations for the NGC. This matrix shows the complex correlation among the attributes. This covariance matrix will be used to assign the error-bars presented in Fig. 3.8 and obtain the χ^2 values presented in Fig. 3.9

3.3.2.2 Angular Cross Correlation

The anisotropies in target quasar density of a pure cosmological signal should not correlate with Galactic foregrounds and survey-related properties. To test this assumption, we measure the cross power spectrum between quasar density and imaging templates. Due

Figure 3.4: Covariance matrix of the mean density contrast for the NGC region estimated from the null EZmock realizations.

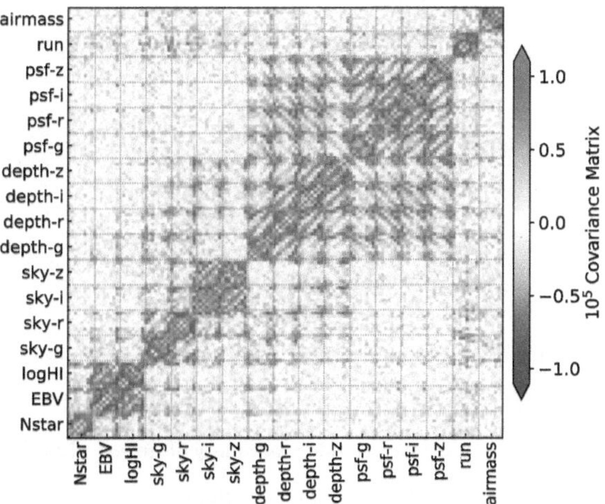

to imaging systematics, the cross power spectrum measurements are not zero, and we use this tool to quantify spurious fluctuations. The cross-spectrum is squared and then normalized by the auto power spectrum of the imaging map,

$$\tilde{C}_\ell^{q,x} = (C_\ell^{q,x})^2 / C_\ell^{x,x}, \tag{3.15}$$

to convert it to the estimated contribution of systematics up to the first order to the auto power spectrum of the quasar density. The auto- or cross-power spectrum is computed over the coefficients $a_{\ell m}$ from spherical harmonic decomposition (e.g., Hobson, 1931),

$$C_\ell^{p,q} = \frac{1}{2\ell + 1} \sum_{m=-\ell}^{\ell} a_{\ell m}^p a_{\ell m}^q, \tag{3.16}$$

with $p = q$ for auto-correlation and $p \neq q$ for cross-correlation. We bin the cross- and auto-correlations, and use only the four lowest ℓ bins centered at $\ell = 1.6, 3.7, 8.3,$ and 18.8, as we are only interested in large-scale modes[28] around $k < 0.01\ h/\mathrm{Mpc}$. Similar to the mean density contrast diagnostic, the covariance matrix C is constructed from the null EZmock cross power spectra, and then unbiased by the Hartlap factor. Finally, we quantify the significance of the residual cross power against the 17 imaging maps by,

$$\chi^2_{\mathrm{tot}} = \tilde{C}^{\dagger}_{\ell} C^{-1} \tilde{C}_{\ell}, \tag{3.17}$$

where,

$$\tilde{C}_{\ell} = [\tilde{C}^{q,x_1}_{\ell}, \tilde{C}^{q,x_2}_{\ell}, ..., \tilde{C}^{q,x_{17}}_{\ell}]. \tag{3.18}$$

3.3.3 Power Spectrum Multipoles

We use the estimator presented in Hand et al. (2017) and implemented in NBODYKIT[29] (Hand et al., 2017) to measure the power spectrum of the DR16 sample for the NGC and SGC regions, separately. We provide a summary of the algorithm below.

We assume a flat ΛCDM cosmology with $\Omega_M = 0.31, h = 0.676, \Omega_b h^2 = 0.022, \sigma_8 = 0.8,$ and $n_s = 0.97$ (Alam et al., 2017) to convert each redshift to distance. After transforming the coordinates, we first begin with digitizing the quasar and random catalogues over a 3D cubic box with a side length of 6600 Mpc/h and 512^3 voxels. Then, we construct the Feldman-Kaiser-Peacock field (FKP; Feldman et al., 1994) and interpolate it using the Triangular Shaped Cloud (TSC) scheme,

$$F(\mathbf{r}) = n_g(\mathbf{r}) - \alpha n_r(\mathbf{r}), \tag{3.19}$$

where $\alpha = \sum n_g(\mathbf{r})/\sum n_r(\mathbf{r})$. The terms n_g and n_r represent the observed density of the quasar sample and random objects weighted by Eq. 3.1, respectively. The power spectrum multipoles are then computed from the FKP field as (Yamamoto et al., 2006),

$$P_{\ell}(k) = \frac{2\ell + 1}{4\pi A} \iiint d\Omega dr_1 dr_2 F(\mathbf{r}_1) F(\mathbf{r}_2) e^{-i\mathbf{k}.(\mathbf{r}_2-\mathbf{r}_1)} \mathcal{L}_{\ell}(\mathbf{k}.\hat{\mathbf{r}}_h) \tag{3.20}$$

---

[28]We use the approximation $\ell + 1/2 \sim k D_A$ with D_A being the angular size distance.

[29]https://nbodykit.readthedocs.io

where $\mathbf{r}_h = (\mathbf{r}_1 + \mathbf{r}_2)/2$ is the line-of-sight distance to the middle point of a given pair, $\mathcal{L}$ is the Legendre polynomial of order ℓ, and $A = \int d\mathbf{r}\bar{n}^2(\mathbf{r})$ is the normalization for the field F, which can be approximated by a discrete sum over the synthetic objects weighted by w_r (see, Eq. 3.1),

$$A = \alpha \sum_i w_r(\mathbf{r}_i)n_r(\mathbf{r}_i). \tag{3.21}$$

Equation 3.20 can be simplified even further. Under the small angle approximation, $\hat{\mathbf{k}}.\hat{\mathbf{r}}_h \sim \hat{\mathbf{k}}.\hat{\mathbf{r}}_2$, the two integrals on r_1 and r_2 are separable. Using the decomposition of the Legendre function into spherical harmonics,

$$\mathcal{L}_\ell(\hat{\mathbf{k}}.\hat{\mathbf{r}}) = \frac{4\pi}{2\ell + 1} \sum_{m=-\ell}^{\ell} Y_{\ell m}(\hat{\mathbf{k}})Y^*_{\ell m}(\hat{\mathbf{r}}), \tag{3.22}$$

we obtain,

$$P_\ell(k) = \frac{2\ell + 1}{A} \int \frac{d\Omega}{4\pi} F_0(\mathbf{k})F_\ell(-\mathbf{k}), \tag{3.23}$$

where

$$F_\ell(\mathbf{k}) = \int d\mathbf{r}F(\mathbf{r})e^{i\mathbf{k}.\mathbf{r}}\mathcal{L}_\ell(\hat{\mathbf{k}}.\hat{\mathbf{r}}) \tag{3.24}$$

$$= \frac{4\pi}{2\ell + 1} \sum_{m=-\ell}^{\ell} Y_{\ell m}(\hat{\mathbf{k}}) \int d\mathbf{r}F(\mathbf{r})e^{i\mathbf{k}.\mathbf{r}}Y^*_{\ell m}(\hat{\mathbf{r}}). \tag{3.25}$$

We measure the power spectrum using Eq. 3.23 in k bands of width $\Delta k = 0.0019\ h\mathrm{Mpc}^{-1}$. This method requires less number of Fourier's transforms compared to other estimators in Bianchi et al. (2015); Scoccimarro (2015). When computing the transforms, the NBDOYKIT software mitigates aliasing with the interlace grid technique presented in Sefusatti et al. (2016) and accounts for any effects caused by the TSC gridding using the factor presented in Jing (2005). Finally, we estimate the shotnoise by summing over the quasar and random catalogue objects,

$$P_{\mathrm{shotnoise}} = \frac{1}{A}[\sum_i w_g^2(\mathbf{r}_i) + \alpha^2 \sum_j w_r^2(\mathbf{r}_j)], \tag{3.26}$$

and subtract it only from the monopole power spectrum ($\ell = 0$).

3.4   Results

This section presents the analyses for the characterization of systematic error and the treatment of spurious fluctuations with various cleaning approaches. We compare the measured power spectra before and after accounting for imaging systematic effects with various linear and nonlinear cleaning methods. In the end, we present the clustering analysis of the EZmock realizations in order to assess the impact of systematic treatments on the measured power spectrum and how one can account for such effect in theoretical modeling.

3.4.1   Method Benchmarks

3.4.1.1   Linear vs Nonlinear Treatment

Our series of tests begin with applying various linear and nonlinear cleaning methods to the main sample of quasars in the NGC region. To conduct a fair comparison, the same set of the four known maps are employed as the input templates. We use the Mean Squared Error and the Poisson Negative Log-Likelihood as two alternatives for the cost function. When using PNLL as the cost function, *Softplus* is applied to the output to satisfy the boundary condition $\lambda > 0$. We also experiment with a nonlinear model without the learning rate annealing to test the sensitivity of training to local minima.

Fig. 3.5 shows the mean density contrast as a function of Galactic extinction in the top panel and the measured power spectrum in the bottom panel, after applying different cleaning methods. The 1σ uncertainty is shown via the grey shades and is constructed from the null EZmock realizations. Linear-MSE here is equivalent to the standard treatment except that the cost function of the former is based on the pixelated density while the latter is based on the binned density. This figure illustrates that our implementations of linear-MSE and linear-PNLL yield similar residual fluctuations to that of the standard treatment. This result implies that with linear regression, adopting the Poisson statistics does not help. Interestingly, the residual fluctuations are reduced substantially after accounting for non-linearities by the hidden layers in the *NN-MSE* and *NN-PNLL* architectures. We also find that turning off the learning rate cycling does not hugely impact the neural network performance (*NN-PNLL-lr*) with marginal effect on the power spectrum, which proves the robustness of the pipeline against local minima and saddle points. This test shows that the most substantial improvement in the measured power spectrum is enabled by accounting for nonlinear systematic effects using the neural network-based methods and then by adopting PNLL in the cost function. Interestingly, our linear cleaning method yields a lower clustering power than the standard linear approach. This test demonstrates that cost function needs to be defined accurately for different levels of model complexity; the choice of PNLL over MSE makes more substantial difference for the NN approach compared with the linear model.

3.4.1.2 Stellar Density from Gaia DR2

The quality of the selection function derived from a template-based cleaning method relies on the available input templates. We may fail to properly eliminate spurious fluctuations if a primary imaging map is missed or our available imaging maps are not completely representing the underlying systematic effects. In this test, we experiment with the available SDSS and non-SDSS imaging maps to find the optimal number of crucial imaging maps, which one would need to explain the residual trends in the mean density contrast. As mentioned before, the standard cleaning approach uses only four maps, i.e., Galactic extinction, depth in g band, sky brightness in i band, and seeing in i band.

Fig. 3.6 top panel shows the mean density contrast against the Gaia stellar density from Gaia Collaboration et al. (2018) after training a neural network with various combinations of the imaging maps using PNLL as the cost function. We also plot the measured density contrast after the standard treatment (*standard*) as our reference for comparison. The

Figure 3.5: *Top*: Mean density contrast of the NGC main sample against Galactic extinction after applying various mitigation techniques, including linear with MSE, linear with PNLL, NN with MSE, NN with PNLL, and NN with PNLL but without learning rate annealing (NN-PNLL-lr). All methods use the *known* set of imaging maps. *Bottom*: Monopole of the NGC main sample after applying the same techniques. The shades represent 1σ statistical uncertainty constructed from the null EZmock realizations.

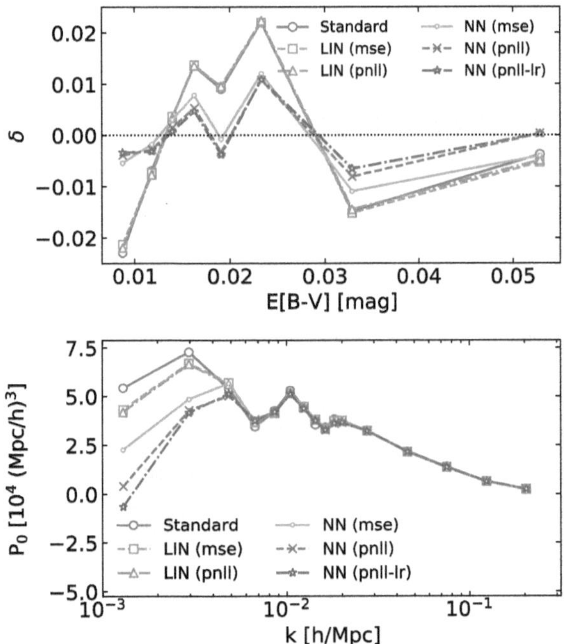

measured monopole power spectrum is shown in the bottom panel. This exercise illustrates that the four known maps are not sufficient to eliminate the systematic trend in the mean density against the Gaia stellar map. We also show that incorporating the SDSS stellar

Figure 3.6: *Top*: Mean density contrast of the NGC main sample against the Gaia

stellar density after accounting for systematics using different combinations of imaging

templates *Bottom*: Monopole power spectrum of the NGC main sample for the same

techniques. The shades represent 1σ statistical uncertainty constructed from the null

EZmock realizations.

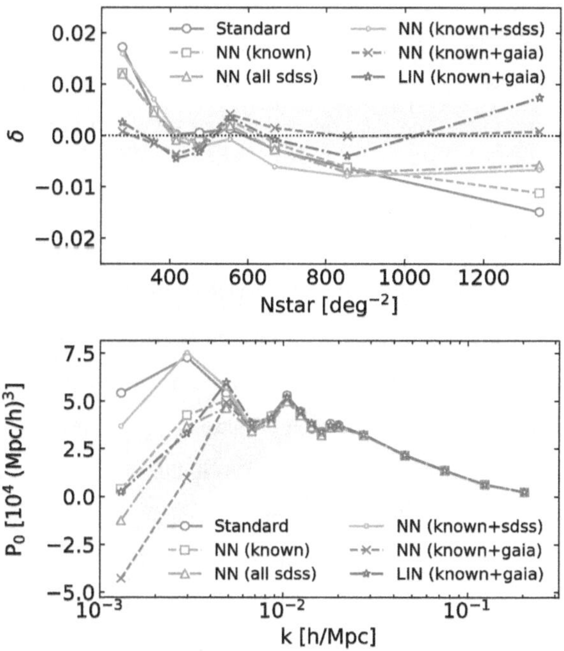

map (*NN known+sdss*) and even all of the SDSS maps (*NN all sdss*) are not adequate to

obtain satisfactory cleaning. On the other hand, we note that *NN all sdss* performs as

well as *Lin Gaia*, implying that the nonlinear nature of NN can mitigate the effect of the

missing input to some extent. Interestingly, adding the SDSS stellar map to the four known maps impacts the result adversely, especially on the low density end of the Gaia DR2 stars. This result might mean that the SDSS stellar density map is not a proper proxy of the stellar contamination effect in the regions with a lower stellar density. This test indicates that the Gaia DR2 stellar map is proved pivotal to perform a robust cleaning of data, that is consistent with the statistical tests of the mocks. For comparison, we apply the linear model with MSE and the additional Gaia map (*LIN known+gaia*), and obtain a total chi-squared value of $\chi^2 = 196.9$ which is still significant (see Table 3.1). This test demonstrates that capabilities of linear treatment is limited even after including the Gaia map, and the nonlinear treatment is crucial, not only when we know the proper set of the input templates a priori, but also when we do not know.

3.4.1.3   Redshift Slicing and Pixel Resolution

The effects of imaging systematics on target density might vary along the line of sight, and alter the slope of the redshift distribution of quasars as well as its overall magnitude. This effect can be ideally investigated by slicing the sample into smaller redshift bins to construct a 3D selection function of imaging systematics. However, further splitting the sample increases the sparsity and makes it difficult to separate the effects of imaging systematics and noise. Thus, we begin with splitting the main sample in the NGC with NSIDE $= 512$ into $0.8 < z < 1.5$ and $1.5 < z < 2.2$ and then we perform regression on each slice separately. We use PNLL as the cost function with the four known maps and the Gaia stellar map as input. Fig. 3.7 compares the mean density contrast as a function of depth in g band (*top*) and the measured monopole power spectrum (*bottom*) after this treatment (*NN 512-2z*) with that of the standard method and the neural network, trained on $0.8 < z < 2.2$ with NSIDE$=512$ (*NN 512-1z*). This plot demonstrates that there is no evidence for redshift-dependent systematic effects due to imaging.

Figure 3.7: *Top*: Mean density contrast of the NGC main sample against depth in the g-band after training the neural network methods with two redshift slices with NSIDE=512 (*NN-512-2z*), one redshift slice with NSIDE=256 (*NN-256-1z*), and one redshift slice with NSIDE=512 (*NN-512-1z*). *Bottom*: Monopole power spectrum after the same techniques. The shades represent 1σ statistical uncertainty constructed from the null EZmock realizations.

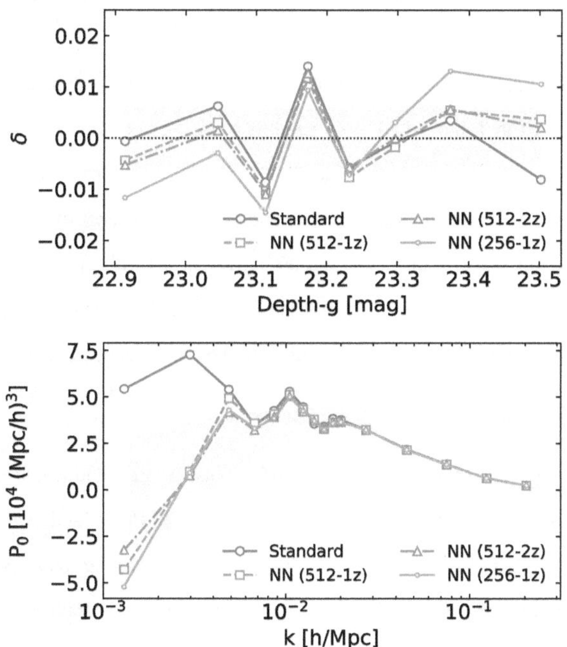

We also train the neural network with coarser imaging templates in NSIDE=256 on $0.8 < z < 2.2$ (*NN 256-1z*). The mean density contrast and measured power spectrum after this treatment are shown in Fig. 3.7 in the top and bottom panels, respectively. The mean density contrast histogram is computed against the depth template in NSIDE=512 and indicates noticeable residual variations from zero in the regions at low and high *depth-g*. This test demonstrates that the systematic weights obtained in the lower resolution cannot properly reduce the trends in the higher resolution. As a caveat, if the top panel was drawn against the depth template in NSIDE=256, we would have gotten a reasonable performance of NN-256. The bottom figure shows the resulting power spectrum accounting for the effects in all 17 imaging maps. We find a reasonable stability against the varied pixel resolutions and choose NN-512 as our default.

The total χ^2 of the mean density residuals for various systematic treatment methods are summarized in Table 3.1. The χ^2 value is 1344.9 before correcting for systematic effects. After using the linear cleaning techniques, the χ^2 value drops below 220. The non-linear approach lowers the error below 200, and changing the cost function from MSE to PNLL improves the performance even further by returning a value around 170, which is less than what is observed in the null EZmock realizations. As a comparison, purely cosmological signal without systematics, estimated from the null EZmocks, returns the χ^2 value of 178.

3.4.2 *Significance of Residual Fluctuations*

For testing the significance of residual systematics, and as our default NN approach, we focus on the neural network method trained with PNLL and cyclic learning rate on the DR16 sample covering $0.8 < z < 2.2$ with the five imaging maps as input features. From the previous analyses, see Table 3.1, we identify this configuration as the optimal approach.

Fig. 3.8 illustrates the observed density contrast of the DR16 quasars as a function of imaging properties for the NGC (top) and SGC (bottom). Respectively from left to right, the imaging quantities are the Gaia DR2 stellar density, Galactic extinction, sky brightness in i band, depth in g band, and seeing in i band. Each panel is annotated with the

Table 3.1: Total χ^2 of the mean density residuals of the main sample in the NGC after various mitigation configurations. The chi-squared value before accounting for imaging systematics is 1344.9, which is much larger than the 95-th quantile observed in the null EZmocks, i.e., 178.

			templates		
NSIDE-*Split*		*known*	*known+SDSS*	*all SDSS*	*known+Gaia*
	standard	218.1	-	-	-
	linear-mse	213.5	-	-	196.9
512-1*z*	linear-pnll	210.2	-	-	-
	nn-mse	194.6	-	-	-
	nn-pnll-lr	**168.99**	-	-	-
	nn-pnll	**163.97**	184.6	**153.9**	**151.7**
512-2*z*	nn-pnll	-	-	-	**165.5**
256-1*z*	nn-pnll	-	-	-	217.6

residual squared errors that are calculated against a zero model as the ground truth[30]. The error bars are obtained from the EZmock catalogues. We observe the biggest variation is against the extinction for about 8% with $\chi^2/\mathrm{dof} = 374.95/8$ in the NGC and 15% against depth-g with $\chi^2/\mathrm{dof} = 828.74/8$ in the SGC. Interestingly, the neural network treatment is capable of modeling and removing the non-linear effects in the sample. On the other hand, the standard treatment leaves a significant chi-squared value against the extinction with $\chi^2/\mathrm{dof} = 37.25/8$ in the NGC.

We compute the total χ^2 of the mean density residuals against all of the 17 imaging maps to determine the significance of spurious fluctuations in the observed mean density of quasars. Fig. 3.9 shows the distributions of χ^2_{tot}, which are constructed from the null (Null) and contaminated EZmocks (Cont), before and after applying imaging systematics

---

[30]We assume that the density contrast must be zero when averaged over many pixels in the absence of imaging systematics.

119

Figure 3.8: Density contrast of the main sample as a function of the primary imaging properties in the NGC (top) and SGC (bottom) before and after accounting for systematic effects using the standard method or neural network. The error-bars are estimated from the null EZmocks and used to calculate the χ^2 of the mean density residuals against each imaging property.

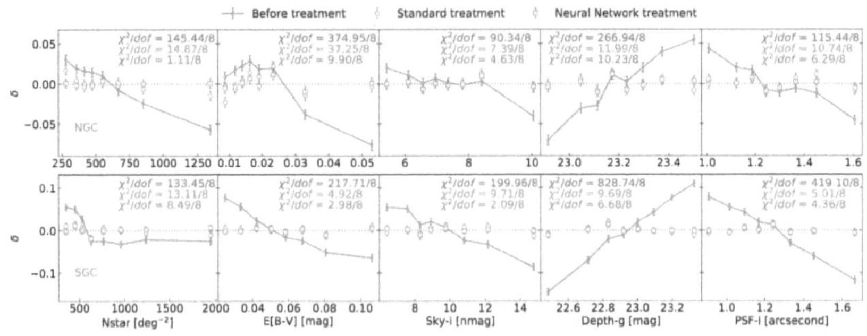

Table 3.2: Total χ^2 of the mean density residuals for the main and high-z samples, before and after mitigation, and the 95-th percentile of null EZmocks. The null EZmock covariance matrix is used to calculate these statistics.

		Noweight	Standard	NN	95-th %
NGC	$0.8 < z < 2.2$	1344.9	218.1	151.7	178.0
	$2.2 < z < 3.5$	1752.0	121.6	104.2	–
SGC	$0.8 < z < 2.2$	1943.0	132.5	116.3	179.2
	$2.2 < z < 3.5$	2553.7	146.0	130.1	–

mitigation for the NGC (left) and SGC (right). The values observed in the DR16 sample before and after cleaning are represented with vertical lines. We use the distribution of the

Figure 3.9: Total χ^2 of the mean density residuals for the EZMocks with systematics (Cont) and without systematics (Null) for the NGC (left) and SGC (right). The same statistics observed in the eBOSS DR16 quasar sample (before and after treatment) are shown via vertical lines with the associated p-values, which are derived by comparing with the mocks without systematics, Null (Truth). There is substantial remaining systematics in the NGC with the standard linear treatment. Also, the contaminated simulations do not reflect the same level of systematic effects as the DR16 sample.

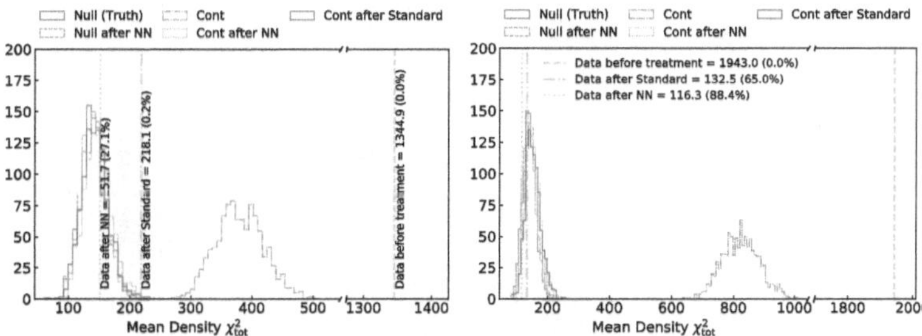

null mocks (Null Truth) to compute $p-$value. In the NGC, the standard treatment yields $\chi^2 = 218.1$ with $p-$value $= 0.2\%$, while the neural network treatment cleans the sample substantially and returns a smaller $\chi2$ and higher $p-$value, respectively, 151.7 and 27.1%. In the SGC, we observe that the standard method returns 132.5 with $p-$value $= 65.0\%$. This residual is somewhat expected since the trends against imaging maps in the SGC are mostly linear, and thus a linear model is sufficient for cleaning (see Fig. 3.8). In the SGC, both methods return statistics that are in agreement with the χ^2 distribution of the null mocks. No remaining systematic error is observed within the statistical uncertainty of the mocks. The total χ^2 of the mean density residuals for the main and high-z samples are reported in Table 3.2. The 95-th percentile for the main sample is estimated from the mocks and reported in the last column. Note that for the NGC region, the standard approach yields a χ^2 value that is larger than the 95-th quantile of the mocks.

Figure 3.10: Similar to Fig. 3.9 for the angular cross power spectrum between the projected quasar density and imaging maps. Compared with the simulations without systematics (Null), there is a substantial remaining systematics both in the NGC and SGC with the standard linear treatment, while the 1D diagnostic (Fig. 3.9) shows no obvious issues with the SGC sample cleaned with the linear approach.

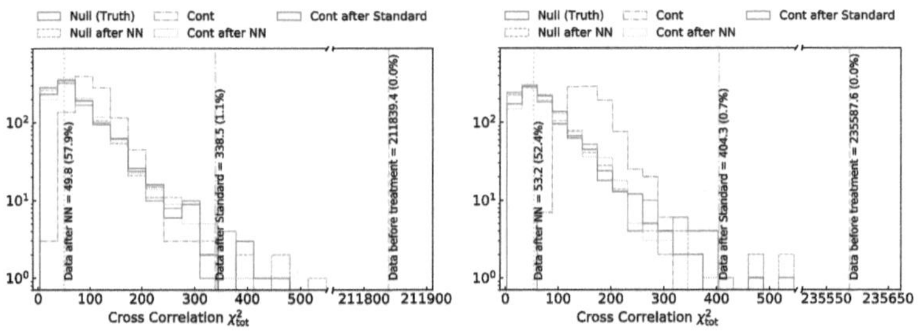

Similarly, we cross-correlate the quasar density map with all of the 17 imaging templates. The cross-correlations are then binned to decrease statistical fluctuations, and normalized by the auto power spectrum of the imaging maps. The first four bins are then used to compute the residual squared error against zero. Fig. 3.10 demonstrates the distribution of χ^2_{tot} constructed from the null and contaminated mocks before and after applying imaging systematics mitigation for the NGC (left) and SGC (right). The values observed in the DR16 sample are represented with vertical lines. In the NGC, the standard treatment returns $\chi^2 = 338.5$ with $p-$value $= 1.1\%$, while the neural network treatment provides a cleaner sample with $\chi2 = 49.8$ and $p-$value $= 57.9$, respectively. In the SGC, we observe that the standard method is incapable of removing the systematics by returning

$\chi^2 = 404.3$ with p–value $= 0.7\%$. On the other hand, the neural network approach enables rigorous cleaning with $\chi^2 = 53.2$ and p–value $= 52.4\%$. This test motivates further investigations of linear systematic treatment methods in future galaxy surveys since the 1D diagnostic based on the mean density contrast is not sufficiently sensitive to unveil these issues with the standard treatment in the SGC (see Fig. 3.9). Interestingly, these histograms show that the magnitude of simulated systematic effects for the contaminated mocks are stronger in the SGC (cf. the left and right panels of Fig. 3.9 and 3.10). Similarly, the DR16 sample shows a stronger spurious fluctuation around 15% against depth in the SGC region, compared with 8% in the NGC.

In summary, we find that the nonlinear aspect of our cleaning approach is the primary reason for efficiently reducing spurious fluctuations and systematic error. The neural network-based approach can model the non-linear feedback of observed quasar density to imaging templates, which results in a cleaner sample with a significantly lower χ^2 value. Although the standard treatment passes the null test based on mean density contrasts, however, the test based on cross power shows that the catalog with the standard systematic weights is not properly cleaned. We find no improvement in the mean density residual after including all SDSS maps for training; however, our analysis shows that the Gaia stellar density is required to satisfy the null tests for residual systematic errors. We also find that computing the mean quasar density per pixel does not change our conclusion based on the mean quasar density per imaging bin, although the former quantity is subject to more fluctuations. Finally, we do not observe a significant change by splitting the main sample into redshift subsamples or using coarser imaging templates. In the following, our neural network-based treatment is applied on the entire $0.8 < z < 2.2$ and uses the Poisson cost function (nn-pnll), cyclic learning rate, and five imaging templates in NSIDE$= 512$ (Known+Gaia) as input features, see, Tab. 3.1.

3.4.3 Power Spectrum

3.4.3.1 Measurement

We now proceed to measure the power spectrum of the DR16 sample and EZmock simulations for each galactic cap separately since each cap is subject to different targeting properties. Fig. 3.11 shows the measured monopole power spectrum P_0 of the main sample in the NGC (left) and SGC (right). We use the square-root of the diagonal terms of the covariance matrices, constructed from the null EZmocks, as the errorbars on P_0. The open and filled circles represent the measured spectrum after cleaning the sample with the standard and neural network treatments, respectively. In both regions, the nonlinear cleaning approach returns a lower power at small k. On the other hand, the effect on the small-scale clustering is very small. We also show various models with $f_{\mathrm{NL}} = -10, 0,$ or 90 to illustrate the sensitivity of the signal on the low-k measurements.

Figure 3.11: Monopole of the main sample in the NGC (left) and SGC (right) after treatment with the standard method and neural network. Various f_{NL} models are plotted to show the sensitivity of the signal on large scales. The shades represent 1σ statistical uncertainty estimated from the EZmocks. The x-axes are logarithmic for $k < 0.02\ h\mathrm{Mpc}^{-1}$ and linear otherwise.

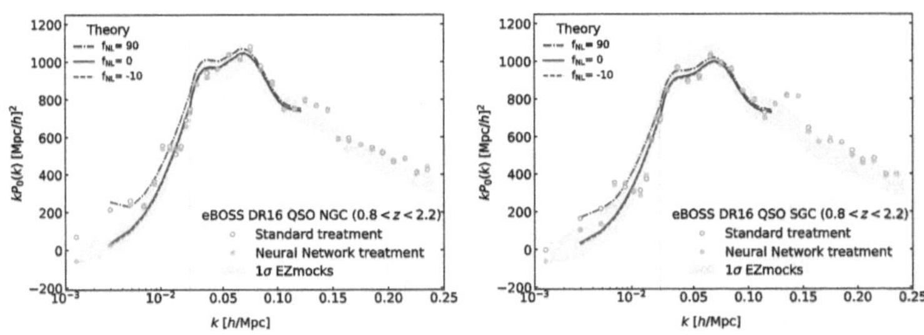

We apply the systematics treatment methods on both the null and contaminated EZmock catalogues to characterize the impact of the mitigation procedure on the measured clustering statistics. The measured power spectrum of the null mocks without any systematic treatment is considered as the ground truth clustering. Fig. 3.12 shows the mean and the standard deviation of the measured spectra from the EZmock realizations in the NGC (left) and SGC (right) regions. The top row illustrates the difference between the mean P_0 of the EZmocks after mitigation, including the null (*Truth after NN*) and contaminated catalogues using the neural network (*Cont after NN* and the standard approach *Cont after Standard*), and the truth clustering (*Truth*). The light and dark shades show the standard deviation of the null EZmock spectra and the 1σ uncertainties on the mean of the mock spectra, respectively. In the second row, we show the relative difference in the mean power spectrum. In the third row, we present the standard deviation of the mock spectra. Finally, we show the relative dispersion in the bottom row. The measured spectra for the contaminated mocks before treatment is an order of magnitude larger than the truth clustering and thus is not visualized for clarity. We note that the magnitude of the excess power observed in the mocks is one order of magnitude smaller than what is observed in the real sample (cf. Fig. 3.11), primarily because a linear model was used to generate systematics. This implies that the actual systematics of the real sample is substantially more severe and complex than in these mocks.

Due to allowing the correction to account for more freedom, the neural-network treatment removes more of the modes, known as the over-fitting problem, when it is applied to the mocks. On the other hand, the standard treatment indicates less of this over-fitting issue. This is expected as the same linear model is used to produce the systematic effects in the mock realizations. The standard deviation of the mock spectra shows that the imaging treatments do not increase the fractional variance of the measured power spectrum down to

$k = 0.003\ h/\text{Mpc}$. Interestingly, we observe that the dispersion of the null mocks decreases after applying the NN treatment, which is due to the over-correction of the power itself [31].

3.4.3.2 Mitigation Bias

Using the mocks, we attempt to estimate any residual or over-correction that might have been introduced in the measured eBOSS QSO power spectrum in the process of the imaging systematics treatment. This assessment is crucial for obtaining an unbiased clustering measurement, which will lead to an accurate inference of cosmological parameters.

We compare the spectra of the contaminated mocks after mitigation to that of the null mocks before treatment as the true power. Fig. 3.12 shows that the NN-based mitigation tends to introduce overcorrection for $k < 0.003$, especially if it is applied on the density field with no systematics (i.e., Truth with NN. [32]). With this caveat, we focus on the overcorrection observed in the contaminated mocks and inspect its nature. The differences between the measured power of the 1000 contaminated mocks after cleaning and the true power for all mocks are shown in Fig. 3.13 as a function of the true power for the first few k bins. We find that the mitigation bias (or overcorrection) is almost linearly proportional to the true power. While this exercise shows that if the data has no systematics or is subject to simple, linear systematics, the neural network based method we developed will potentially introduce a small degree of overcorrection and we can attempt to correct for such mitigation bias. However, from Fig. 3.9 and 3.10, it is apparent that the DR16 sample is subject to much more severe and nonlinear systematic effects compared to the mocks and the standard linear method is not sufficiently effective. Therefore, we believe that the overcorrection is

---

[31]The error on the power spectrum is expected to be proportional to the power itself under a Gaussian limit.

[32]The standard method performs better by construction, as it assumes we know exactly the source of systematics.

less likely a problem for the real eBOSS data and an attempt to mitigate it might further bias the data. We present the discussion of our mitigation bias on primordial non-Gaussianity constraints in a companion paper (Mueller and et al. in prep., 2020).

3.5 Conclusion

We have performed a thorough study of imaging systematic effects and various template-based mitigation techniques in the final sample of quasars (Lyke et al., 2020; Ross et al., 2020) from the eBOSS DR16 (Ahumada et al., 2020). We present a nonlinear cleaning approach, based on artificial neural networks, and compare the treatment effectiveness with the standard method, based on linear regression. The methods are applied to model the observed density of quasars given a set of templates for imaging properties, related to SDSS properties and Galactic foregrounds, which include stellar density, Galactic extinction, neutral hydrogen column density, depth, seeing, sky brightness, airmass, and run. As summarized in Tab. 3.1,

1. We find that the neural network-based approach outperforms standard linear regression by allowing more freedom for correcting nonlinear and complex variations in the quasar density caused by imaging properties, see Fig. 3.5 and 3.8. The approach is also further improved by using the Poisson statistics to account for the sparsity of the DR16 sample.

2. Stellar density is one of the most important sources of spurious fluctuations, and a new template constructed using the Gaia DR2 (Gaia Collaboration et al., 2018) yields the best agreement to the observed chi-squared values in the simulations, see Fig. 3.6 and Tab. 3.1. We also show that linear treatment, with the Gaia map included, is still not able to properly remove systematics.

3. We find no evidence for redshift-dependent imaging systematics and no substantial difference after changing the pixel resolution of imaging templates, see Fig. 3.7. Therefore we choose NN trained with PNLL, cyclic learning, and imaging templates in NSIDE = 512 as our default approach.

We utilize the EZmocks, both in the presence and absence of imaging systematics, to construct covariance matrices, quantify residual systematic error, and assess the quality of the DR16 sample for cosmological studies. We find

1. The mean density null test shows some remaining systematic error in the catalogue with the standard weights in the NGC region, specifically $\chi^2 = 218.1$ with $p-$value = 0.2%. Although this test does not reveal any issues with the standard catalogue in the SGC region, $\chi^2 = 132.5$ with $p-$value = 65.0% (see Tab. 3.2), our second null test based on cross-power spectra unveils a significant systematic error in the SGC, $\chi^2 = 404.3$ with $p-$value = 0.7%.

2. This work motivates further investigations of linear systematic treatment methods in future galaxy surveys since the 1D diagnostic based on the mean density contrast does not indicate any issues with the standard treatment in the SGC (see Fig. 3.9).

3. The catalogue with the neural network-based systematic weights passes both null tests by providing substantially lower χ^2 values, see Fig. 3.9 and 3.10.

Collectively, these tests demonstrate that the DR16 quasar catalogue with the standard systematic weights suffer from residual imaging systematics in both Galactic caps, and should not be used for measuring quasar clustering on large scales, i.e., $k < 0.01\ h/\mathrm{Mpc}$, as shown in Fig. 3.11. Nevertheless, it is expected that the impact of imaging systematics to be insignificant on the BAO measurements (e.g., analyses presented in Hou et al., 2021; Neveux et al., 2020), and a thorough investigation is conducted in a companion paper (Merz et al., 2021).

We then apply our methods on the EZmocks to quantify the impact of systematics treatment on quasar clustering measurements, see Fig. 3.12. The neural-network treatment removes some of the cosmological power due to allowing for more freedom in removing systematic effects. We find that the impacts of overfitting on the mean of the mock power spectra and its error are marginal for $k > 0.004\ h/\mathrm{Mpc}$. We employ linear regression to model the impact of mitigation on recovering the ground truth clustering, see Fig. 3.13. We emphasize that the utility of the mitigation bias treatment is not clear since the parameters are derived from the mocks without realistic imaging systematics, see Fig. 3.9 and 3.10. However, we investigate the effect on primordial non-Gaussianity constraints in a companion paper (Mueller and et al. in prep., 2020).

The end-product from this work is a new value-added quasar catalogue with the enhanced weights to correct for nonlinear imaging systematic effects. The new weights are necessary to make a robust measurement of quasar clustering on large scales ($k < 0.01\ h/\mathrm{Mpc}$). This catalogue is used in a companion paper constraining the local-type primordial non-Gaussianity (Mueller and et al. in prep., 2020).

Figure 3.12: Measured power spectrum of the EZmock realizations before and after systematic treatment for the NGC (left) and SGC (right) regions. From top to bottom, we show the difference between the mean mitigated spectrum and the mean truth spectrum, the relative difference, the dispersion in the mock spectra, and the relative dispersion.

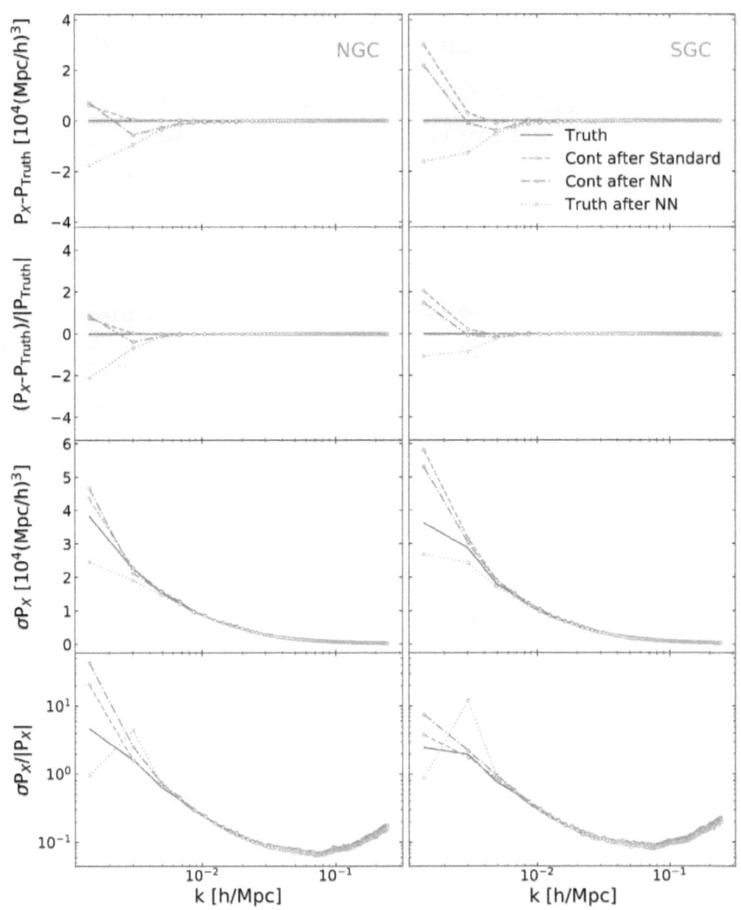

Figure 3.13: Difference between the measured spectra of the contaminated EZmock catalogues after cleaning and the spectra of the null catalogues as a function of the latter. The medians are used to obtain the best linear fit in each k bin, and are shown only for $k = 0.001\ h/\mathrm{Mpc}$ with open squares.

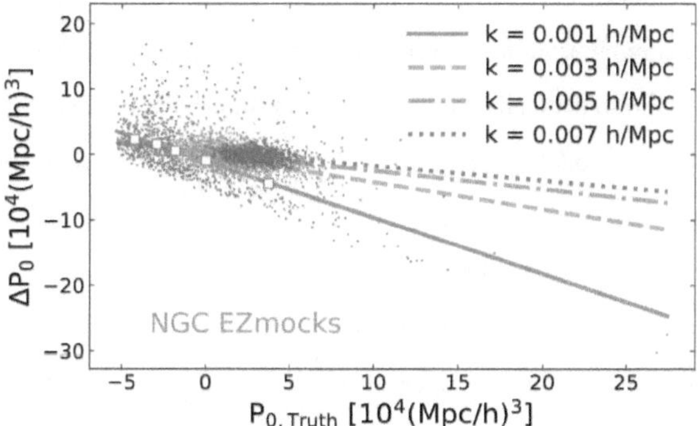

4 Local Primordial Non-Gaussianity with Imaging Data

This Chapter uses galaxy and quasar samples selected from the DESI Legacy Imaging Surveys Data Release 9 to constrain $f_{\rm NL}$ using the scale-dependent effect on the large-scale clustering of biased tracers. This data release offers more than 14000 $\deg^2$ sky coverage which is expected to yield very promising results, e.g., $\sigma(f_{\rm NL}) = 10$ (Anna Porredon, personal communication). Nevertheless, this type of primordial non-Gaussianity study so far has been dramatically limited by severe signatures of imaging systematics on large scales. The methodology presented in Chapter 2 and enhanced in 3 allow me to rigorously clean and prepare data for such a complicated and sensitive analysis. I use simulations to validate modeling pipeline, evaluate data quality, and conduct statistical tests to assess whether or not spurious fluctuations are properly mitigated for a r obust $f_{\rm NL}$ constraint.

Unfortunately the catalogue cleaning and preparation of the DR9 samples are not complete at the time of writing this book. Therefore, I only present the result for the mock test. This chapter is structured as follows. Section 4.1 outlines the theoretical power spectrum which includes redshift space distortions and local primordial non-Gaussianity. Section 4.3 presents simulations based on Gaussian density fluctuations. In Section 4.4, I present the 1D posterior of $f_{\rm NL}$ from analyzing the mean power spectrum of the simulated realizations, and finally conclude this chapter in Section 4.5

4.1 Modeling Angular Power Spectrum

The angular power spectrum of galaxy number counts with redshift space distortions is related to the linear power spectrum $P(k)$ by (Padmanabhan et al., 2007; Slosar et al., 2008),

$$C_\ell = \frac{2}{\pi} \int_0^\infty \frac{dk}{k} k^3 P(k) |\Delta_\ell(k)|^2, \tag{4.1}$$

where $\Delta_\ell(k) = \Delta_\ell^{\mathrm{g}}(k) + \Delta_\ell^{\mathrm{RSD}}(k) + \Delta_\ell^{f_{\mathrm{NL}}}(k)$ with,

$$\Delta_\ell^{\mathrm{g}}(k) = \int \frac{dr}{r} rb(r)D(r)\frac{dN}{dr} j_\ell(kr), \tag{4.2}$$

$$\Delta_\ell^{\mathrm{RSD}}(k) = -\int \frac{dr}{r} rf(r)D(r)\frac{dN}{dr} j_\ell''(kr), \tag{4.3}$$

$$\Delta_\ell^{f_{\mathrm{NL}}}(k) = f_{\mathrm{NL}}\frac{\alpha}{k^2 T(k)} \int \frac{dr}{r} r[b(r) - p]\frac{dN}{dr} j_\ell(kr), \tag{4.4}$$

where $\alpha = 3\delta_c\Omega_M(H_0/c)^2$, $b(r)$ is the halo bias, dN/dr is the normalized redshift distribution of galaxies[33], $D(r)$ is the growth factor scaled to one at $r = 0$, $f(r)$ is the growth rate, and r is the comoving distance. The parameter p is the response of the tracer to halo's gravitational field, e.g., 1.6 for recent mergers. I employ the FFTLog[34] algorithm and its extension presented in Fang et al. (2020) to compute the inner integrations over $d\ln r$ and Simpson's rule to evaluate the outer integration over $d\ln k$.

4.2 Modeling Survey Geometry Effect

For a galaxy survey that observes the sky partially, the measured power spectrum is convolved with the survey geometry. This means that the pseudo-power spectrum $\hat{C}_\ell$ obtained by the direct Spherical Harmonic Transforms of a partial sky map, differs from the full-sky angular spectrum C_ℓ. However, their ensemble average is related by (Hivon et al., 2002)

$$< \hat{C}_\ell >= \sum_{\ell'} M_{\ell\ell'} < C_{\ell'} >, \tag{4.5}$$

where $M_{\ell\ell'}$ represents the mode-mode coupling from the partial sky coverage. This is known as the Window Function effect and a proper assessment of this effect is crucial for a robust measurement of the large-scale clustering of galaxies. This window effect is a source

---

[33] $dN/dr = (dN/dz) * (dz/dr) \propto (dN/dz) * H(z)$

[34] github.com/xfangcosmo/FFTLog-and-beyond

of observational systematic error and impacts the measured galaxy clustering, especially on large scales ($\ell < 200$).

We follow a similar approach to that of (Chon et al., 2004) to model the window function effect on the theoretical power spectrum C_ℓ rather than correcting the measured pseudo-power spectrum from data. First, we use HEALPIX to compute the pseudo-power spectrum of the window, which is defined by a mask file in ring ordering format with NSIDE= 256. Then, we transform it to correlation function by,

$$\omega^{\text{window}}(\theta) = \frac{1}{4\pi} \sum_\ell (2\ell + 1)\hat{C}_\ell^{\text{window}} P_\ell(\cos\theta). \tag{4.6}$$

Next, we normalize ω^{window} such that it is normalized to one at $\theta = 0$. Finally, we multiply the theory correlation function by ω^{window}, and transform the result back to ℓ-space[35],

$$C_\ell^{\text{model}} = 2\pi \int d\theta\, \omega(\theta) P_\ell(\cos\theta). \tag{4.7}$$

4.3 Simulations

We simulate Gaussian mocks with a constant bias $b = 1.5$ and redshift distribution defined as,

$$n(z) \propto \frac{1}{2z_0}\left(\frac{z}{z_0}\right)^2 \exp(-z/z_0), \tag{4.8}$$

where $z_0 = 0.0417 i_{\text{lim}} - 0.744$ with $i_{\text{lim}} = 26.0$, as shown in Fig. 4.1. First, we compute the theoretical power spectrum C_ℓ from Eq. 4.1 with $f_{\text{NL}} = 0$, and then generate Gaussian randoms $a_{\ell m}$, with a zero mean and variance equal to C_ℓ, up to $\ell = 499$ to construct a density contrast δ,

$$\delta = \sum_\ell \sum_{m=-\ell}^{\ell} a_{\ell m} Y_{\ell m}(\theta, \phi), \tag{4.9}$$

---

[35]We use Gauss-Legendre Quadrature to do the integration over $d\theta$.

where $Y_{\ell m}$ represents the spherical harmonic function of degree ℓ and order m. Fig. 4.2 shows a realization of the full sky Gaussian mocks and its corresponding partial sky mock.

Figure 4.1: Bias and redshift distribution of Gaussian simulations as a function of redshift.

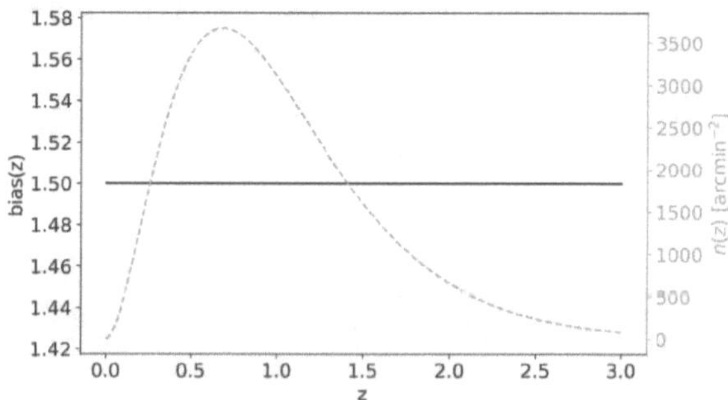

Figure 4.2: Mollweide projection of full sky (left) and partial sky (right) Gaussian mocks.

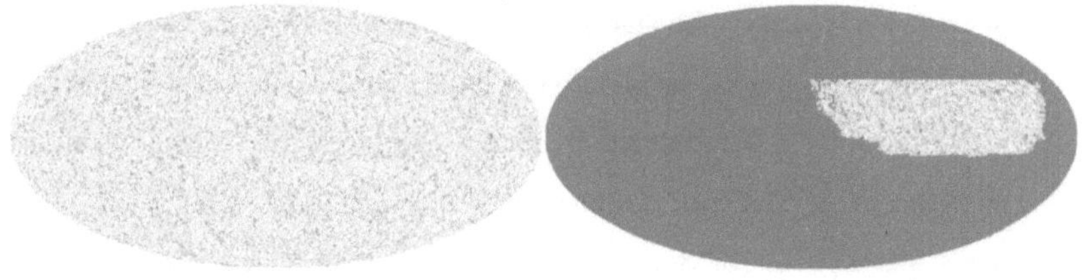

We generate 2000 realizations for the full sky and 2000 for the partial sky case, and apply HEALPIX software to measure power spectrum from each realization. Fig. 4.4 shows the mean power spectrum of the full sky mocks and that of the partial sky mocks along with theoretical predictions (with $f_{NL} = 0$). We also show two cosmologies with non-zero PNG, $f_{NL} = 10$ and -10, with dotted curves to show that the impact of survey geometry is comparable to that of PNG, and therefore for a robust analysis we should account for survey geometry.

Figure 4.3: Window correlation function of the DECaLS North footprint.

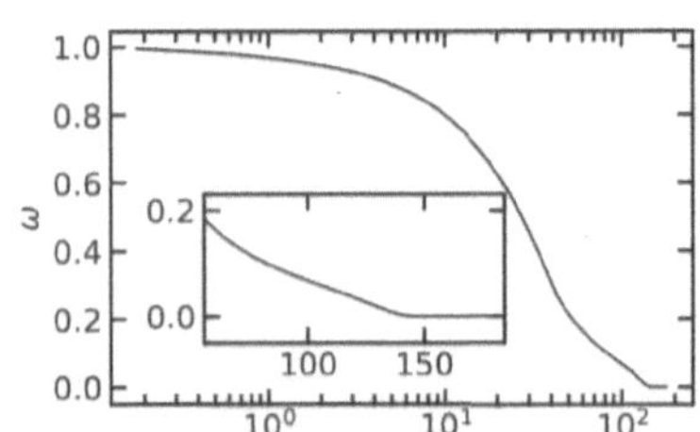

4.4   1D Posterior

I now proceed to apply the Markov chain Monte Carlo (MCMC) regression on the mean power spectrum of the full sky mocks. I use a Gaussian likelihood, with covariance matrix from the mocks, and a flat prior on the non-Gaussianity parameter f_{NL} between -100 and 100. Fig. 4.5 illustrates the mean power spectrum of the full sky mocks and that of the partial-sky mocks along with the theoretical predictions with various f_{NL} values. For the partial-sky case, we show the window-convolved theoretical models as well with blue curves. As shown in Fig. 4.6, my MCMC is able to recover the truth f_{NL} value, i.e., zero, within the one σ confidence interval, i.e., $-42.3 < f_{NL} < 5.7$. Interestingly, an asymmetry is noticed in the posterior of f_{NL}, which can be associated with the fact that negative values

Figure 4.4: The mean measured power spectrum of the Gaussian mocks and theoretical predictions for the full- and partial-sky realizations. The contours represent 1σ C.L. constructed from the partial sky mocks.

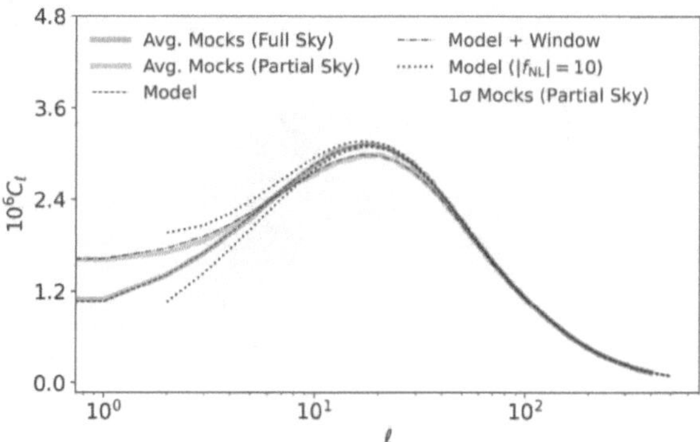

of f_{NL} are more likely to be in agreement with the mock power spectrum, as shown in Fig. 4.5.

Figure 4.5: Mean measured power spectrum of the full-sky (partial-sky) mocks in the left (right) and theoretical predictions for cosmologies with various f_{NL} values. The errorbars are obtained from the standard deviation of the mock spectra.

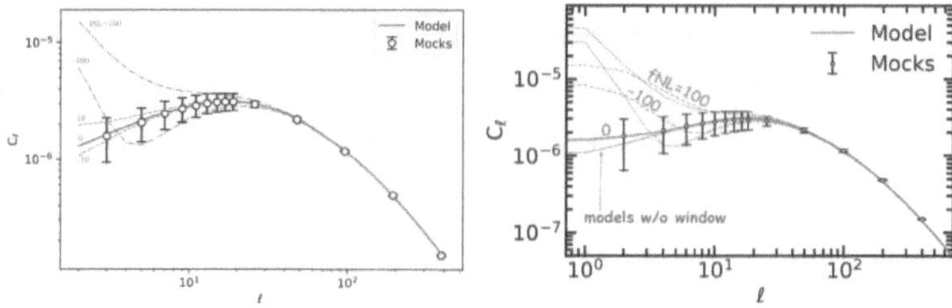

Figure 4.6: 1D posterior of f_{NL} from the full-sky (blue) and partial-sky (red) Gaussian mocks. The mean posterior is shown with a vertical line.

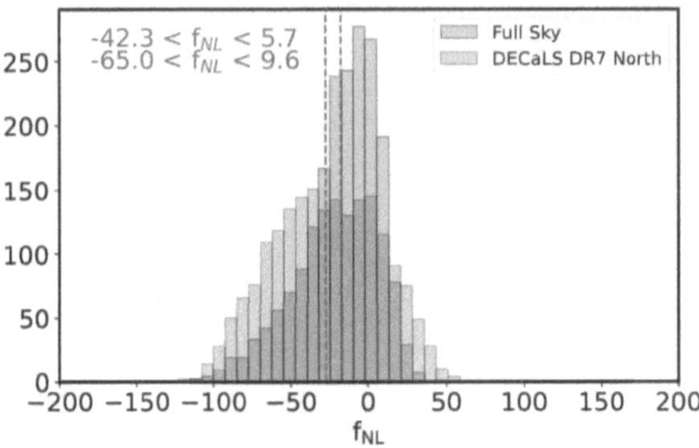

4.5 Conclusions and Future Outlook

I have presented an analysis of the local primordial non-Gaussianity using simulations with $f_{\mathrm{NL}} = 0$. I demonstrate that the pipeline performs robustly on the full-sky and partial-sky mocks and that the true signal is recovered within 1σ confidence interval. In future works, this analysis will be applied to the recent release of DESI imaging data. One of the primary challenges is to ensure that the real DESI imaging samples are sufficiently cleaned. Also, the mocks used here do not have the same redshift distribution and clustering signal as the real data. Thus, I envision that more realistic mock datasets are needed to construct covariance matrices and test for remaining systematic error. Note that these mocks need to be analyzed separately for the BASS/MzLS, DECaLS North and DECaLS South regions since each footprint might have a different targeting efficiency and depth variation.

5 Broader Impact

5.1 Primordial Non-Gaussianity with EBOSS Quasars

Led by Dr. Eva Mueller, this project utilizes the eBOSS quasar catalogue, developed in Chapter 3, to constrain primordial non-Gaussianity (PNG) parameter f_{NL}, and will be published in Mueller et al. (in prep.).

The quasar sample from eBOSS occupies the largest spectroscopic volume up to date, spanning a redshift range of $0.8 < z < 2.2$. This large survey volume allows the measurement of the power spectrum at very large scales to an unprecedented precision. To measure quasar clustering on large scales robustly, one must account for the evolution of bias over redshift by properly weighting each object. As derived in Mueller et al. (2019), we apply redshift weights designed to minimize the uncertainty in f_{NL} to fully exploit all the information in the LSS of the quasar clustering. The redshift weighting technique was first developed for BAO measurements (Zhu et al., 2015, 2016) and was subsequently applied to optimize RSD measurements (Ruggeri et al., 2017).

While the FKP weights balance the shot noise and cosmic variance contributions to minimize statisical uncertainty, the redshift weights account for the redshift evolution of the underlying physical theory to further improve the constraints. This methodology does not require binning the sample in redshift space, and it reduces the information loss at the boundaries of the redshift bins. In this analysis, we only focus on the power spectrum monopole and thus the redshift weight is defined as,

$$w_z(z) = (b + \frac{1}{3}f)(b - p)D(z), \tag{5.1}$$

where f is the growth rate, D is the growth function, and the fiducial bias b is given by $b(z) = 0.53 + 0.29(1 + z)^2$. We apply these weights to each object in the data and randoms. The total weight is then

$$w_{tot} = w_{FKP}\, w_{CP}\, w_{NOZ}\, w_{sys}\, \sqrt{|w_z|}. \tag{5.2}$$

With this weighting scheme, the effective redshift of $z_{\mathrm{eff}} = 1.51$ increases to $z_{\mathrm{eff}} = 1.80$, reflecting the up-weighting to a higher redshift due to the redshift weights.

Because f_{NL} signal is not sensitive to small scale clustering, where nonlinear physics becomes important, our analysis here is restricted to linear scales with the linear redshift power spectrum $P(k, \mu)$ modelled as,

$$P(k,\mu) = \frac{(b + \Delta b + \mu^2 f)^2 P_m(k)}{1 + (k\mu\sigma_v/H_0)^2}, \tag{5.3}$$

with the matter power spectrum $P_m(k)$, the quasar bias b, scale-dependent PNG biasing effect Δb, the growth rate f, and a Lorenzian damping factor given the velocity dispersion σ_v. The power spectrum multipoles can then be calculated as Eq. 1.19,

$$P_\ell(k) = \frac{2\ell + 1}{2} \int_{-1}^{1} d\mu P(k,\mu)\mathcal{L}_\ell(\mu).$$

The constraints on f_{NL} along with the nuisance parameters b and σ_v are then inferred from optimizing the Gaussian likelihood $\mathcal{L}$ using a Monte Carlo Markov Chain (MCMC) Metropolis-Hasting algorithm,

$$-2\ln\mathcal{L} = \chi^2,$$
$$= (D - Y(p))^\dagger C^{-1}(D - Y(p)), \tag{5.4}$$

where D refers to the vector of measured power spectrum, C is the covariance matrix from 1000 EZmocks, $Y(p)$ is the theory prediction of power spectrum, and p represents the set of parameters, i.e., $p = \{f_{\mathrm{NL}}, \sigma_v, b\}$. We use flat priors on all parameters, and apply the Gelman-Rubin diagnostic to assure the convergence of the MCMC chains (Gelman et al., 1992; Brooks and Gelman, 1998). We obtain that all MCMC chains yield a convergence criteria of $R < 0.01$ or lower. In our analysis we marginalize over the shot noise, the linear bias, and the velocity dispersion σ_v as nuisance parameters. With other cosmological parameters fixed, we vary only four parameters in total in our MCMC runs. Fig. 5.1

shows the 1D posterior on f_{NL} for the combined sample (left) and each galactic cap separately (right) with the standard treatment (solid) and neural network approach (dashed). Interestingly, this work illustrates the importance of imaging systematics and how they can bias the inference of cosmological parameters.

Figure 5.1: 1D posterior on f_{NL} using the catalogs with the default treatment (solid) and neural network (dashed), for both galactic caps combined (left) and each cap separately (right). *Source*: Mueller and et al. in prep. (2020).

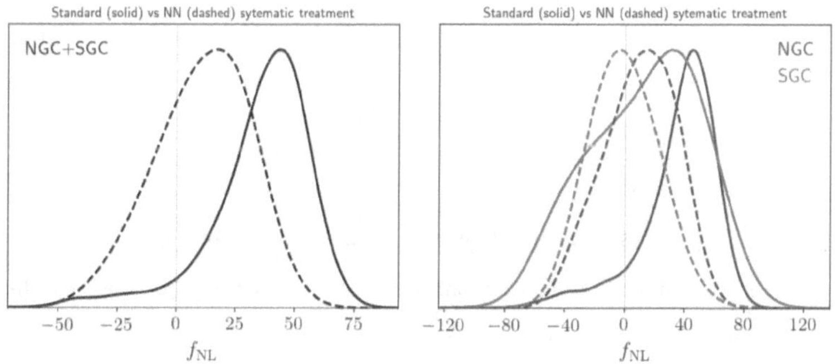

In Fig. 5.2, we show the history of f_{NL} constraints from CMB and LSS surveys from the 2000s to 2021. These constraints are expected to improve with the next generation of galaxy surveys such as DESI and Rubin Observatory, down to a level that is competitive with CMB measurements, e.g., $\sigma(f_{NL}) \sim 5$. Given the precision of future surveys, handling non-linear systematic effects will be more important than ever.

5.2 Baryon Acoustic Oscillations in Projected Cross-Correlation

Led by Dr. Pauline Zarrouk and published in Zarrouk et al. (2021), this work attempts to measure BAO parameters and cosmological distances in the cross-correlation

Figure 5.2: History of f_{NL} constraints from CMB and LSS surveys and their cross correlation study. *Source*: Mueller and et al. in prep. (2020).

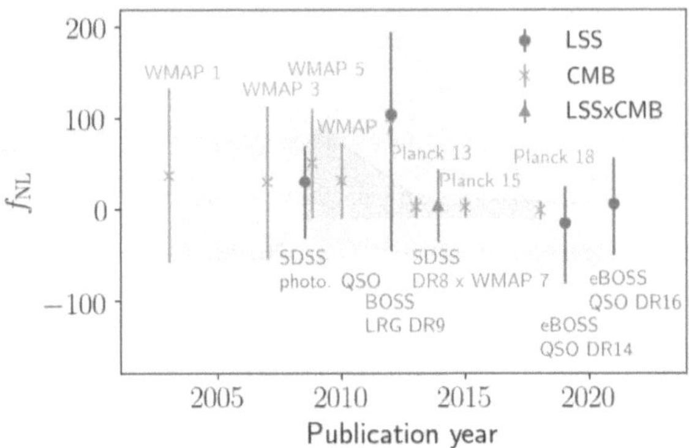

of dense galaxies from DESI imaging and sparse quasars from eBOSS spectroscopy. The methodology presented in Chapter 2 is employed for the systematic treatment and preparation of the DESI imaging data.

Measurement of BAO scale in the large-scale clustering of galaxies and quasar at different redshifts is a well-established technique for constraining dark energy parameters. However, statistics at high redshifts are extremely hindered by fainter and sparser targets. Cross correlating different tracers of dark matter is forecast to improve statistical confidence levels by canceling the sample variance and breaking the degeneracy between tracer bias and other cosmological parameters.

The default DESI ELG selection does not provide sufficient targets for this analysis, with a mean target density of 2400 deg^{-2} while we would require a mean density of 3000 deg^{-2} to achieve a signal to noise ratio around five. We select less star-forming objects (i.e., with redder colors) to reach a higher redshift density, especially since we are

Figure 5.3: *Left*: Color-color diagram showing the selection used in this analysis (red), the DESI ELG main selection (black), and the eBOSS ELG selection (magenta). The color-codding represents photometric redshifts from HSC-PDR2. *Right*: Likelihood of the transverse BAO parameter α_{cross} in terms of $\Delta\chi^2$ in the redshift range $0.6 \leq z \leq 1.2$. The solid curves display the likelihood obtained when fitting the data with a model that contains the BAO feature, while the dashed curves display the same information for a model without BAO. *Source*: Zarrouk et al. (2021).

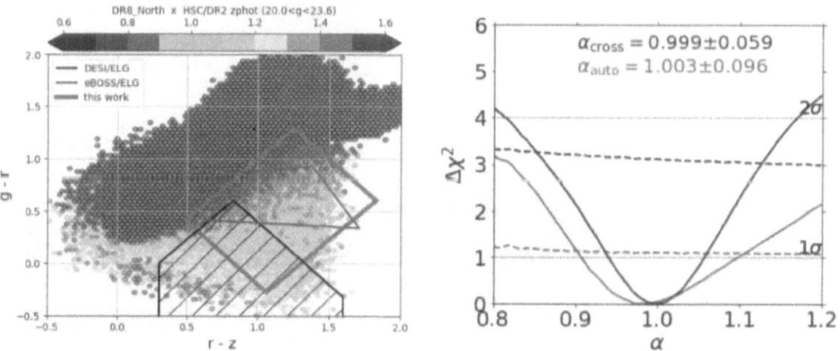

not limited by desiring a minimal [OII] flux. Fig. 5.3 shows the $g - r$ vs $r - z$ color-color diagram for galaxies in the g-band magnitude range with our selection (red box), the DESI ELG selection (black box) and the eBOSS ELG selection (magenta box). The color-coding shows the photometric redshifts of the galaxies after matching the ELG targets with HSC-PDR2 (Aihara et al., 2019). Since we need to select fainter objects at the limit of the survey depth, the photometric sample is therefore expected to be more prone to density fluctuations due to inhomogeneities in the selection. To address this issue, I have applied the neural network technique from Chapter 2 to the photometric sample and confirmed that it can correct for complicated variations that standard multivariate linear regression techniques cannot.

We perform two analyses in parallel; one from the auto-correlation function of the eBOSS DR16 quasars and one from the projected cross-correlation function, both under the same fitting procedure. Given the low statistical precision of the quasar sample in the redshift ranges we consider, it is more optimal to fit an isotropic shift α_{iso} and constrain the spherically-averaged distance D_{V}:

$$D_{\mathrm{V}} = \left[(1 + z)^2 cz \frac{D_{\mathrm{M}}^2}{H} \right]^{\frac{1}{3}},$$ (5.5)

where c is the speed of light and α_{iso} is thus defined by:

$$\alpha_{\mathrm{iso}} = \frac{D_{\mathrm{V}}(z) r_{\mathrm{drag}}^{\mathrm{fid}}}{D_{\mathrm{V}}^{\mathrm{fid}}(z) r_{\mathrm{drag}}},$$ (5.6)

where r_{drag} is the sound horizon at the drag epoch and corresponds to the expected location of the BAO feature in comoving distance units, and the superscript 'fid' represents the values of a fiducial cosmology. The auto-correlation function enables the constraint of $D_{\mathrm{V}}(z)$ by fitting an isotropic shift α_{iso} while the projected cross-correlation function puts constraints on the comoving angular diameter distance $D_{\mathrm{M}}(z)$ by fitting a transverse shift $\alpha_{\perp}$,

$$\alpha_{\perp} = \frac{D_{\mathrm{M}}(z) r_{\mathrm{drag}}^{\mathrm{fid}}}{D_{\mathrm{M}}^{\mathrm{fid}}(z) r_{\mathrm{drag}}}.$$ (5.7)

We find that the cross-correlation technique can reduce shot noise, and thus provide better constraints on the cosmic distance (6% precision) compared to the results obtained from the auto-correlation (9% precision) as show in the right panel Fig. 5.3. We also find that we are limited by the number density and purity of the photometric sample and its overlap in redshift with the spectroscopic sample, which thus affects the performance of the method. Nevertheless, the technique developed in this work will be even more promising with the arrival of deeper photometric data thanks to upcoming surveys such as *Euclid* (Amendola et al., 2013). By performing a proper comparison between the two statistics, this study highlights the utility of the cross-correlation between photometry and spectroscopy for robust BAO measurements. The method could be applied to cross correlating DESI quasars, which will still be limited by shot noise (Aghamousa et al., 2016), with a sample of Hα (0.7 $< z <$ 2 and [OIII] emission galaxies (2 $< z <$ 2.7) selected from *Euclid* (Mehta et al., 2015). This technique should enable better constraints on the transverse BAO scale at $z \geq 2$ than DESI will with the auto-correlation function of the quasars alone.

5.3 Testing Imaging Systematics on Baryon Acoustic Oscillation

This work is led by Grant Merz, and will be published in Merz et al. (2021). My role was to prepare the real data and simulated catalogs, help with the pipeline development, and contribute to the manuscript.

BAO feature is considered as a robust standard ruler against various theoretical and observational systematic effects. The current consensus is that spurious fluctuations, caused by varying imaging properties, have a negligible impact on the typical scale of baryon acoustic oscillations. For instance, Ross et al. (2017b) performed a thorough analysis of the observational systematic effects on the BAO measurement of the BOSS DR12 massive galaxies in the redshift interval of 0.2 $< z <$ 0.75. They characterized a systematic trend in galaxy density against stellar density and seeing, and mitigated using linear regression. They showed that the effect on the BAO fit before and after the mitigation was less than 10-15% of the quoted statistical precision of 1.3% and 2.4%. Neveux et al. (2020) performed the BAO analysis with the final sample of quasars (Ross et al., 2020) from the eBOSS DR16 (Ahumada et al., 2020), and found that the change in the BAO measurement before and after the mitigation is about 30% of the statistical precision. This impact is still reasonable when accounting for the statistical scatter due to a stronger observational systematic effects of the eBOSS DR16 sample compared to that of the BOSS data.

However, Chapter 3 shows that the default treatment still leaves a spurious clustering signal on very large scales in the power spectrum of the eBOSS DR16 QSO sample and that a new set of weights is required to account for nonlinear imaging systematics. This particular form of nonlinear systematics has not ever been investigated for BAO measurements to date. It is therefore of paramount importance to assess whether or not these new systematic weights are still required for the BAO measurement, and to quantify any impacts on the statistical uncertainty. This project employs the new value-added quasar catalogs from Chapter 3 for testing the robustness of the BAO measurement presented in Neveux et al. (2020) and for quantifying the effects of imaging systematics when the scatter in the treatment process is propagated to the final constraint. We also investigate additional freedom in the BAO fitting compared to the default used in Neveux et al. (2020) and characterize the interplay between imaging systematics and the flexibility of the BAO fitting model.

The BAO signal is isolated by subtracting a smooth fit power spectrum from the best-fit spectrum. Figure 5.4 shows the BAO signal in the NGC for eBOSS catalogs mitigated using NN weights derived from various combinations of imaging templates. Subtracted from both the data and model is a smooth power spectrum. Error bars are taken from the covariance matrix of 1000 EZmocks. For visualization clarity, the models and data points are shifted horizontally. The quadrupole (and any higher order spherical harmonics) signal is in general noisier than the monopole, and makes the BAO signal harder to isolate on its own, which can be seen in Figure 5.4. However, it is important to simultaneously fit the multipoles using the full covariance matrix to extract the complete BAO signal. We find that the α values derived from the fits to NN-weighted data are lower than those found from fitting the default-weighted data. The parameter α_{iso} appears to be more stable across the different fits than the anisotropic $\alpha_{\parallel}$ and $\alpha_{\perp}$. The error ellipses corresponding to different fits are shown in Figure 5.5, with 68% and 95% confidence intervals. Both the default fit and NN fits have in general around a 12-15% increase in error compared to the result from Neveux et al. (2020). We note this is most likely due to our error being derived from the posterior of the Markov Chains. All fits with the NN-based catalogs having a slightly larger error than the fit with the default catalog.

Overall, we find that the α parameters do not differ significantly across the fits on the measured spectra after the various treatments. Changes in best-fit $\alpha_\perp$ and $\alpha_\parallel$ are within the error from the posterior. This result implies the systematic weights derived from the NN approach does not significantly alter the BAO shape or locations of the BAO peaks compared to the linear regression method, and supports the robustness of the eBOSS BAO result against observational systematics. However, these changes are significant in the context of future surveys such as DESI, which aim for very precise measurements with deeper and wider coverage, and thus it will be crucial to fully investigate different systematic error mitigation methods in order to produce accurate measurements within such a small precision.

Figure 5.4: *Left*: The isolated BAO wiggles in the power spectra of the eBOSS quasars in the NGC. Points are the data with error bars derived from the covariance matrix and the solid lines are the best-fit models. The wiggles are isolated by subtracting the power spectra by a smoothed spectra, which is produced from fitting the default catalog to a smoothed model. Data points are shifted horizontally to prevent overlap, but best-fits are not shifted. Colors correspond to the catalogs with different systematic weights: blue, dark purple, maroon, magenta and pink showing the default, known with 1 redshift split, known with 2 redshift splits, all with 1 redshift split and all with 2 redshift splits, respectively. *Source*: Merz et al. (2021).

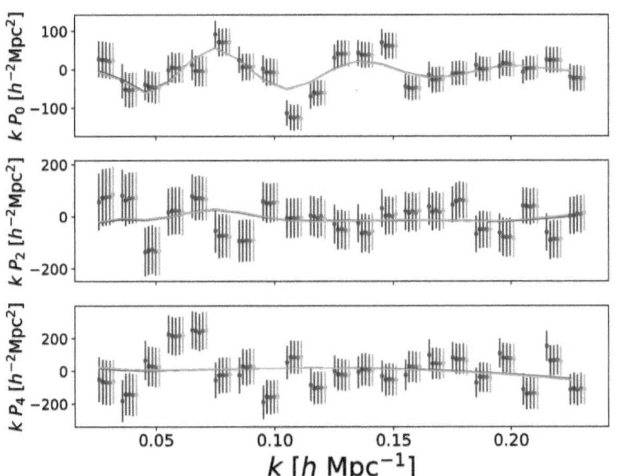

Figure 5.5: Error ellipses at 1 and 2σ confidence for the eBOSS quasar sample in the NGC. The colors are blue (default), red (known;1-zsplit), green (known;2-zsplits), violet (all;1-zsplit) and gold (all;2-zsplits). *Source*: Merz et al. (2021).

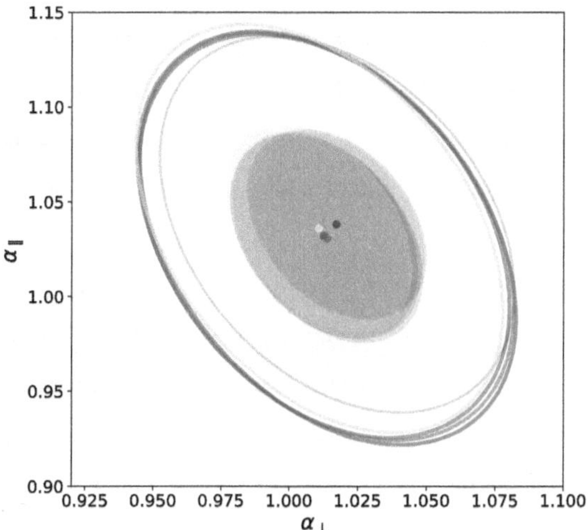

6   Conclusions

The research carried out for this book has established a novel approach based on machine learning and statistics for the characterization and mitigation of systematic errors in galaxy survey data. One primary source of systematic uncertainty is related to imaging properties including but not limited to galactic extinction, local stellar contamination, and seeing. The variation of photometric properties across the survey footprint creates spurious fluctuations in the observed target density field. Many of these effects have large-scale variations, and thus they result in an excess clustering signal at large scales. Combined with other observational systematics, these realities limit our ability to derive robust cosmological constraints.

Chapter 2 presents a rigorous application of a methodology based on artificial neural networks for the treatment of imaging systematics in clustering measurements of galaxies from DECaLS DR7 imaging. The DR7 sample exhibits a strong dependence on Galactic foregrounds, which implies the significance of accurate stellar and Galactic dust maps for the next generation of galaxy surveys. The deep learning-based methodology models the effect of imaging properties on the target density field without making any prior assumption of the linearity of the fitting m odel. I d emonstrate t hat t his a pproach o ffers a higher flexibility t han l inear r egression, a nd t hus i t e nables c apturing t he n onlinear systematic effect on the galaxy density field. A r ecursive f eature s election i s i mplemented and applied to both the DR7 sample and simulations to characterize the primary sources of systematic uncertainties. The feature selection procedure identifies G alactic foregrounds as the primary sources of systematic error in DR7. This is in agreement with other diagnostics based on one-point and 1D statistics, respectively, Pearson's correlation coefficient and mean density contrast. When applied to simulated datasets, I find that in the absence of contamination, the neural network is the most robust approach against regressing out the cosmological clustering. This is mainly due to the feature selection process that appropriately reduces the flexibility of the model. This chapter indicates that any cosmological study that uses DESI imaging will require a thorough analysis of imaging systematics to determine an estimate of the residual systematic uncertainty for the parameters of interest. To use this DR7 sample for cosmological exploitation, I suggest that one should account for 19% additional systematic errors to the statistical error in the density field level. My analysis also shows that additional masking, especially based on depth-g, or improving the methodology to deal with depth and sky background issues would be the next steps to prepare this data for a cosmological analysis. This is of vital importance for DESI, which will use imaging from the same survey for targeting.

Chapter 3 enhances the neural network methodology presented in Chapter 2 to be applicable to sparse targets. For a sample with low surface density, the number count of targets in each pixel follows the Poisson distribution, and the Mean Squared Error may not be the optimal cost function to train neural network parameters. To address this challenge, I implement the Poisson negative log likelihood as the loss function to improve the training procedure. Also, a cyclic learning rate is incorporated to improve the robustness of the training procedure against local minima and saddle points. The methodology is applied to the final sample of quasars from the eBOSS DR16 and realistic simulations. This work substantially improves quasar clustering compared to the conventional cleaning approach, and enables a robust measurement of primordial non-Gaussianity with quasar clustering. We find that local stellar density is one of the most important sources of spurious fluctuations, and a new template constructed using the Gaia DR2 (Gaia Collaboration et al., 2018) yields the best agreement with the observed quasar clustering. I find no evidence for redshift-dependent imaging systematics and no substantial difference after changing the pixel resolution of imaging templates. However, a thorough analysis of imaging systematics along the line-of-sight is of paramount importance for the next generation of galaxy surveys that probe bigger sky coverage and fainter targets near the detection limits.

In Chapter 4, I present the pipeline to analyze DESI imaging, cleaned by the methodology presented in this book, to constrain the local primordial non-Gaussianity. I create Gaussian simulations to test my analysis pipeline and theoretical modeling, and to construct covariance matrices. I run Monte-Carlo Markov Chains, and find that the posterior of the f_{NL} parameter is inherently asymmetric, even though the true f_{NL} is recovered within the 1σ confidence interval. In the immediate future, this analysis will be extended to take into account the survey geometry effect and will be applied to DESI imaging data. Chapter 5 briefly discusses three publications (one peer-reviewed and two in the final stage of preparation), to which I contributed as the second author. These projects are a testament to the impact of the tools and methods developed in this book for tackling fundamental physics questions with galaxy survey data. The results show that careful treatment of imaging systematics will be very crucial for constraining physics of the early universe, but BAO measurements are not significantly affected.

To conclude, future galaxy surveys will provide massive torrents of data to unveil many unknowns in the Universe, ranging from inflation t o n eutrino m asses. The enormously increased data volume will reduce statistical uncertainties significantly, which will enable addressing fundamental questions in cosmology and differentiating between various theoretical models; however, higher precision will demand careful analyses to control and deal with the sophisticated and entangled nature of observational systematic error. Fluctuations in imaging properties and error in photometric calibrations pose as primary challenges in the era of DESI and Rubin Observatory, and potentially limit the extend by which we can exploit the clustering of large-scale structure for cosmology. The techniques and tools presented in this book will be instrumental to improve data quality, activate robust clustering measurements, and enable unbiased cosmological inference.

References

Aartsen, M., Abraham, K., Ackermann, M., Adams, J., Aguilar, J., Ahlers, M., Ahrens, M., Altmann, D., Andeen, K., Anderson, T., et al. (2016). Searches for sterile neutrinos with the icecube detector. *Physical review letters*, 117(7):071801.

Abazajian, K., Adelman-McCarthy, J. K., Agüeros, M. A., Allam, S. S., Anderson, S. F., Annis, J., Bahcall, N. A., Baldry, I. K., Bastian, S., Berlind, A., et al. (2003). The first data release of the sloan digital sky survey. *The Astronomical Journal*, 126(4):2081.

Abazajian, K. N., Adshead, P., Ahmed, Z., Allen, S. W., Alonso, D., Arnold, K. S., Baccigalupi, C., Bartlett, J. G., Battaglia, N., Benson, B. A., et al. (2016). Cmb-s4 science book. *arXiv preprint arXiv:1610.02743*.

Abolfathi, B., Aguado, D., Aguilar, G., Prieto, C. A., Almeida, A., Ananna, T. T., Anders, F., Anderson, S. F., Andrews, B. H., Anguiano, B., et al. (2018). The fourteenth data release of the sloan digital sky survey: First spectroscopic data from the extended baryon oscillation spectroscopic survey and from the second phase of the apache point observatory galactic evolution experiment. *The Astrophysical Journal Supplement Series*, 235(2):42.

Acquaviva, V., Bartolo, N., Matarrese, S., and Riotto, A. (2003). Gauge-invariant second-order perturbations and non-gaussianity from inflation. *Nuclear Physics B*, 667(1-2):119–148.

Ade, P. A., Aghanim, N., Arnaud, M., Ashdown, M., Aumont, J., Baccigalupi, C., Banday, A., Barreiro, R., Bartlett, J., Bartolo, N., et al. (2016). Planck 2015 results-xiii. cosmological parameters. *Astronomy & Astrophysics*, 594:A13.

Adelman-McCarthy, J. K., Agüeros, M. A., Allam, S. S., Anderson, K. S., Anderson, S. F., Annis, J., Bahcall, N. A., Bailer-Jones, C. A., Baldry, I. K., Barentine, J., et al. (2007). The fifth data release of the sloan digital sky survey. *The Astrophysical Journal Supplement Series*, 172(2):634.

Agarwal, N., Ho, S., and Shandera, S. (2014). Constraining the initial conditions of the universe using large scale structure. *Journal of Cosmology and Astroparticle Physics*, 2014(02):038.

Aghamousa, A., Aguilar, J., Ahlen, S., Alam, S., Allen, L. E., Prieto, C. A., Annis, J., Bailey, S., Balland, C., Ballester, O., et al. (2016). The desi experiment part i: Science, targeting, and survey design. *arXiv preprint arXiv:1611.00036*.

Ahmad, Q. R., Allen, R., Andersen, T., Anglin, J., Barton, J., Beier, E., Bercovitch, M., Bigu, J., Biller, S., Black, R., et al. (2002). Direct evidence for neutrino flavor transformation from neutral-current interactions in the sudbury neutrino observatory. *Physical review letters*, 89(1):011301.

Ahmad, Q. R., Allen, R., Andersen, T., Anglin, J., Bühler, G., Barton, J., Beier, E., Bercovitch, M., Bigu, J., Biller, S., et al. (2001). Measurement of the rate of $v_e + d \rightarrow p + p + e^-$ interactions produced by ^{8}b solar neutrinos at the sudbury neutrino observatory. *Physical Review Letters*, 87(7):071301.

Ahn, C. P., Alexandroff, R., Prieto, C. A., Anderson, S. F., Anderton, T., Andrews, B. H., Aubourg, É., Bailey, S., Balbinot, E., Barnes, R., et al. (2012). The ninth data release of the sloan digital sky survey: first spectroscopic data from the sdss-iii baryon oscillation spectroscopic survey. *The Astrophysical Journal Supplement Series*, 203(2):21.

Ahumada, R., Allende Prieto, C., Almeida, A., Anders, F., Anderson, S. F., Andrews, B. H., Anguiano, B., Arcodia, R., Armengaud, E., Aubert, M., Avila, S., Avila-Reese, V., Badenes, C., Balland, C., Barger, K., Barrera-Ballesteros, J. K., Basu, S., Bautista, J., Beaton, R. L., Beers, T. C., Benavides, B. I. T., Bender, C. F., Bernardi, M., Bershady, M., Beutler, F., Bidin, C. M., Bird, J., Bizyaev, D., Blanc, G. A., Blanton, M. R., Boquien, M., Borissova, J., Bovy, J., Brandt, W. N., Brinkmann, J., Brownstein, J. R., Bundy, K., Bureau, M., Burgasser, A., Burtin, E., Cano-Díaz, M., Capasso, R., Cappellari, M., Carrera, R., Chabanier, S., Chaplin, W., Chapman, M., Cherinka, B., Chiappini, C., Doohyun Choi, P., Chojnowski, S. D., Chung, H., Clerc, N., Coffey, D., Comerford, J. M., Comparat, J., da Costa, L., Cousinou, M.-C., Covey, K., Crane, J. D., Cunha, K., da Silva Ilha, G., Dai, Y. S., Damsted, S. B., Darling, J., Davidson, James W., J., Davies, R., Dawson, K., De, N., de la Macorra, A., De Lee, N., de Andrade Queiroz, A. B., Deconto Machado, A., de la Torre, S., Dell'Agli, F., du Mas des Bourboux, H., Diamond-Stanic, A. M., Dillon, S., Donor, J., Drory, N., Duckworth, C., Dwelly, T., Ebelke, G., Eftekharzadeh, S., Eigenbrot, A. D., Elsworth, Y. P., Eracleous, M., Erfanianfar, G., Escoffier, S., Fan, X., Farr, E., Fernández-Trincado, J. G., Feuillet, D., Finoguenov, A., Fofie, P., Fraser-McKelvie, A., Frinchaboy, P. M., Fromenteau, S., Fu, H., Galbany, L., Garcia, R. A., García-Hernández, D. A., Garma Oehmichen, L. A., Ge, J., Geimba Maia, M. A., Geisler, D., Gelfand, J., Goddy, J., Gonzalez-Perez, V., Grabowski, K., Green, P., Grier, C. J., Guo, H., Guy, J., Harding, P., Hasselquist, S., Hawken, A. J., Hayes, C. R., Hearty, F., Hekker, S., Hogg, D. W., Holtzman, J. A., Horta, D., Hou, J., Hsieh, B.-C., Huber, D., Hunt, J. A. S., Ider Chitham, J., Imig, J., Jaber, M., Jimenez Angel, C. E., Johnson, J. A., Jones, A. M., Jönsson, H., Jullo, E., Kim, Y., Kinemuchi, K., Kirkpatrick, Charles C., I., Kite, G. W., Klaene, M., Kneib, J.-P., Kollmeier, J. A., Kong, H., Kounkel, M., Krishnarao, D., Lacerna, I., Lan, T.-W., Lane, R. R., Law, D. R., Le Goff, J.-M., Leung, H. W., Lewis, H., Li, C., Lian, J., Lin, L., Long, D., Longa-Peña, P., Lundgren, B., Lyke, B. W., Ted Mackereth, J., MacLeod, C. L., Majewski, S. R., Manchado, A., Maraston, C., Martini, P., Masseron, T., Masters, K. L., Mathur, S., McDermid, R. M., Merloni, A., Merrifield, M., Mészáros, S., Miglio, A., Minniti, D., Minsley, R., Miyaji, T., Mohammad, F. G., Mosser, B., Mueller, E.-M., Muna, D., Muñoz-Gutiérrez, A., Myers, A. D., Nadathur, S., Nair, P., Nandra, K., do Nascimento, J. C., Nevin, R. J., Newman, J. A., Nidever, D. L., Nitschelm, C., Noterdaeme, P., O'Connell, J. E., Olmstead, M. D., Oravetz, D., Oravetz, A., Osorio, Y.,

Pace, Z. J., Padilla, N., Palanque-Delabrouille, N., Palicio, P. A., Pan, H.-A., Pan, K., Parker, J., Paviot, R., Peirani, S., Peña Ramfez, K., Penny, S., Percival, W. J., Perez-Fournon, I., Pérez-Ràfols, I., Petitjean, P., Pieri, M. M., Pinsonneault, M., Poovelil, V. J., Povick, J. T., Prakash, A., Price-Whelan, A. M., Raddick, M. J., Raichoor, A., Ray, A., Rembold, S. B., Rezaie, M., Riffel, R. A., Riffel, R., Rix, H.-W., Robin, A. C., Roman-Lopes, A., Román-Zúñiga, C., Rose, B., Ross, A. J., Rossi, G., Rowlands, K., Rubin, K. H. R., Salvato, M., Sánchez, A. G., Sánchez-Menguiano, L., Sánchez-Gallego, J. R., Sayres, C., Schaefer, A., Schiavon, R. P., Schimoia, J. S., Schlafly, E., Schlegel, D., Schneider, D. P., Schultheis, M., Schwope, A., Seo, H.-J., Serenelli, A., Shafieloo, A., Shamsi, S. J., Shao, Z., Shen, S., Shetrone, M., Shirley, R., Silva Aguirre, V., Simon, J. D., Skrutskie, M. F., Slosar, A., Smethurst, R., Sobeck, J., Sodi, B. C., Souto, D., Stark, D. V., Stassun, K. G., Steinmetz, M., Stello, D., Stermer, J., Storchi-Bergmann, T., Streblyanska, A., Stringfellow, G. S., Stutz, A., Suárez, G., Sun, J., Taghizadeh-Popp, M., Talbot, M. S., Tayar, J., Thakar, A. R., Theriault, R., Thomas, D., Thomas, Z. C., Tinker, J., Tojeiro, R., Toledo, H. H., Tremonti, C. A., Troup, N. W., Tuttle, S., Unda-Sanzana, E., Valentini, M., Vargas-González, J., Vargas-Magaña, M., Vázquez-Mata, J. A., Vivek, M., Wake, D., Wang, Y., Weaver, B. A., Weijmans, A.-M., Wild, V., Wilson, J. C., Wilson, R. F., Wolthuis, N., Wood-Vasey, W. M., Yan, R., Yang, M., Yèche, C., Zamora, O., Zarrouk, P., Zasowski, G., Zhang, K., Zhao, C., Zhao, G., Zheng, Z., Zheng, Z., Zhu, G., and Zou, H. (2020). The 16th Data Release of the Sloan Digital Sky Surveys: First Release from the APOGEE-2 Southern Survey and Full Release of eBOSS Spectra. *ApJs*, 249(1):3.

Aihara, H., AlSayyad, Y., Ando, M., Armstrong, R., Bosch, J., Egami, E., Furusawa, H., Furusawa, J., Goulding, A., Harikane, Y., Hikage, C., Ho, P. T. P., Hsieh, B.-C., Huang, S., Ikeda, H., Imanishi, M., Ito, K., Iwata, I., Jaelani, A. T., Kakuma, R., Kawana, K., Kikuta, S., Kobayashi, U., Koike, M., Komiyama, Y., Li, X., Liang, Y., Lin, Y.-T., Luo, W., Lupton, R., Lust, N. B., MacArthur, L. A., Matsuoka, Y., Mineo, S., Miyatake, H., Miyazaki, S., More, S., Murata, R., Namiki, S. V., Nishizawa, A. J., Oguri, M., Okabe, N., Okamoto, S., Okura, Y., Ono, Y., Onodera, M., Onoue, M., Osato, K., Ouchi, M., Shibuya, T., Strauss, M. A., Sugiyama, N., Suto, Y., Takada, M., Takagi, Y., Takata, T., Takita, S., Tanaka, M., Terai, T., Toba, Y., Uchiyama, H., Utsumi, Y., Wang, S.-Y., Wang, W., and Yamada, Y. (2019). Second data release of the Hyper Suprime-Cam Subaru Strategic Program. *PASJ*, 71(6):114.

Akrami, Y., Arroja, F., Ashdown, M., Aumont, J., Baccigalupi, C., Ballardini, M., Banday, A., Barreiro, R., Bartolo, N., Basak, S., et al. (2018). Planck 2018 results. i. overview and the cosmological legacy of planck. *arXiv preprint arXiv:1807.06205*.

Akrami, Y., Arroja, F., Ashdown, M., Aumont, J., Baccigalupi, C., Ballardini, M., Banday, A., Barreiro, R., Bartolo, N., Basak, S., et al. (2019). Planck 2018 results. ix. constraints on primordial non-gaussianity. *arXiv preprint arXiv:1905.05697*.

Alam, S., Ata, M., Bailey, S., Beutler, F., Bizyaev, D., Blazek, J. A., Bolton, A. S., Brownstein, J. R., Burden, A., Chuang, C.-H., et al. (2017). The clustering of galaxies in the completed sdss-iii baryon oscillation spectroscopic survey: cosmological analysis of the dr12 galaxy sample. *Monthly Notices of the Royal Astronomical Society*, 470(3):2617–2652.

Alam, S., Aubert, M., Avila, S., Balland, C., Bautista, J. E., Bershady, M. A., Bizyaev, D., Blanton, M. R., Bolton, A. S., Bovy, J., et al. (2020). The completed sdss-iv extended baryon oscillation spectroscopic survey: cosmological implications from two decades of spectroscopic surveys at the apache point observatory. *arXiv preprint arXiv:2007.08991*.

Alcock, C. and Paczyński, B. (1979). An evolution free test for non-zero cosmological constant. *Nature*, 281(5730):358–359.

Allen, T., Grinstein, B., and Wise, M. B. (1987). Non-gaussian density perturbations in inflationary cosmologies. *Physics Letters B*, 197(1-2):66–70.

Alpher, R. A., Bethe, H., and Gamow, G. (1948a). The origin of chemical elements. *Physical Review*, 73(7):803.

Alpher, R. A. and Herman, R. (1948). Evolution of the universe. *Nature*, 162(4124):774–775.

Alpher, R. A., Herman, R., and Gamow, G. (1948b). Thermonuclear reactions in the expanding universe. *Physical Review*, 74(9):1198.

Alvarez, M., Baldauf, T., Bond, J. R., Dalal, N., de Putter, R., Doré, O., Green, D., Hirata, C., Huang, Z., Huterer, D., et al. (2014). Testing inflation with large scale structure: connecting hopes with reality. *arXiv preprint arXiv:1412.4671*.

Amendola, L., Appleby, S., Bacon, D., Baker, T., Baldi, M., Bartolo, N., Blanchard, A., Bonvin, C., Borgani, S., Branchini, E., Burrage, C., Camera, S., Carbone, C., Casarini, L., Cropper, M., de Rham, C., Di Porto, C., Ealet, A., Ferreira, P. G., Finelli, F., García-Bellido, J., Giannantonio, T., Guzzo, L., Heavens, A., Heisenberg, L., Heymans, C., Hoekstra, H., Hollenstein, L., Holmes, R., Horst, O., Jahnke, K., Kitching, T. D., Koivisto, T., Kunz, M., La Vacca, G., March, M., Majerotto, E., Markovic, K., Marsh, D., Marulli, F., Massey, R., Mellier, Y., Mota, D. F., Nunes, N. J., Percival, W., Pettorino, V., Porciani, C., Quercellini, C., Read, J., Rinaldi, M., Sapone, D., Scaramella, R., Skordis, C., Simpson, F., Taylor, A., Thomas, S., Trotta, R., Verde, L., Vernizzi, F., Vollmer, A., Wang, Y., Weller, J., and Zlosnik, T. (2013). Cosmology and Fundamental Physics with the Euclid Satellite. *Living Reviews in Relativity*, 16:6.

Anderson, L., Aubourg, E., Bailey, S., Beutler, F., Bhardwaj, V., Blanton, M., Bolton, A. S., Brinkmann, J., Brownstein, J. R., Burden, A., et al. (2014a). The clustering of galaxies in the sdss-iii baryon oscillation spectroscopic survey: baryon acoustic oscillations in the data releases 10 and 11 galaxy samples. *Monthly Notices of the Royal Astronomical Society*, 441(1):24–62.

Anderson, L. et al. (2014b). The clustering of galaxies in the SDSS-III Baryon Oscillation Spectroscopic Survey: baryon acoustic oscillations in the Data Releases 10 and 11 Galaxy samples. *Mon. Not. Roy. Astron. Soc.*, 441(1):24–62.

Bardeen, J. M., Steinhardt, P. J., and Turner, M. S. (1983). Spontaneous creation of almost scale-free density perturbations in an inflationary universe. *Physical Review D*, 28(4):679.

Bashinsky, S. and Bertschinger, E. (2001). Position-space description of the cosmic microwave background and its temperature correlation function. *Physical review letters*, 87(8):081301.

Baumann, D., Jackson, M. G., Adshead, P., Amblard, A., Ashoorioon, A., Bartolo, N., Bean, R., Beltrán, M., De Bernardis, F., Bird, S., et al. (2009). Probing inflation with cmb polarization. In *AIP Conference Proceedings*, volume 1141, pages 10–120. American Institute of Physics.

Bautista, J. E., Guy, J., Rich, J., Blomqvist, M., Des Bourboux, H. D. M., Pieri, M. M., Font-Ribera, A., Bailey, S., Delubac, T., Kirkby, D., et al. (2017). Measurement of baryon acoustic oscillation correlations at z= 2.3 with sdss dr12 lyα-forests. *Astronomy & Astrophysics*, 603:A12.

Bautista, J. E., Vargas-Magaña, M., Dawson, K. S., Percival, W. J., Brinkmann, J., Brownstein, J., Camacho, B., Comparat, J., Gil-Marín, H., Mueller, E.-M., Newman, J. A., Prakash, A., Ross, A. J., Schneider, D. P., Seo, H.-J., Tinker, J., Tojeiro, R., Zhai, Z., and Zhao, G.-B. (2018). The SDSS-IV Extended Baryon Oscillation Spectroscopic Survey: Baryon Acoustic Oscillations at Redshift of 0.72 with the DR14 Luminous Red Galaxy Sample. *ApJ*, 863(1):110.

Bautista, J. E., Vargas-Magaña, M., Dawson, K. S., Percival, W. J., Brinkmann, J., Brownstein, J., Camacho, B., Comparat, J., Gil-Marín, H., Mueller, E.-M., et al. (2018). The sdss-iv extended baryon oscillation spectroscopic survey: baryon acoustic oscillations at redshift of 0.72 with the dr14 luminous red galaxy sample. *The Astrophysical Journal*, 863(1):110.

Bekhti, N. B., Flöer, L., Keller, R., Kerp, J., Lenz, D., Winkel, B., Bailin, J., Calabretta, M., Dedes, L., Ford, H., et al. (2016). Hi4pi: a full-sky h i survey based on ebhis and gass. *Astronomy & Astrophysics*, 594:A116.

Bergé, J., Gamper, L., Réfrégier, A., and Amara, A. (2013). An ultra fast image generator (ufig) for wide-field astronomy. *Astronomy and Computing*, 1:23–32.

Bernardeau, F. and Uzan, J.-P. (2002). Non-gaussianity in multifield inflation. *Physical Review D*, 66(10):103506.

Bianchi, D., Gil-Marín, H., Ruggeri, R., and Percival, W. J. (2015). Measuring line-of-sight-dependent Fourier-space clustering using FFTs. *MNRAS*, 453(1):L11–L15.

Bianchi, D. and Percival, W. J. (2017). Unbiased clustering estimation in the presence of missing observations. *Monthly Notices of the Royal Astronomical Society*, 472(1):1106–1118.

Blanton, M. R., Bershady, M. A., Abolfathi, B., Albareti, F. D., Prieto, C. A., Almeida, A., Alonso-García, J., Anders, F., Anderson, S. F., Andrews, B., et al. (2017). Sloan digital sky survey iv: Mapping the milky way, nearby galaxies, and the distant universe. *The Astronomical Journal*, 154(1):28.

Blomqvist, M., du Mas des Bourboux, H., Busca, N. G., de Sainte Agathe, V., Rich, J., Balland, C., Bautista, J. E., Dawson, K., Font-Ribera, A., Guy, J., Le Goff, J.-M., Palanque-Delabrouille, N., Percival, W. J., Pérez-Ràfols, I., Pieri, M. M., Schneider, D. P., Slosar, A., and Yèche, C. (2019). Baryon acoustic oscillations from the cross-correlation of Lyα absorption and quasars in eBOSS DR14. *AAP*, 629:A86.

Bolton, A. S., Schlegel, D. J., Aubourg, É., Bailey, S., Bhardwaj, V., Brownstein, J. R., Burles, S., Chen, Y.-M., Dawson, K., Eisenstein, D. J., et al. (2012). Spectral classification and redshift measurement for the sdss-iii baryon oscillation spectroscopic survey. *The Astronomical Journal*, 144(5):144.

Bondi, H. and Gold, T. (1948). The Steady-State Theory of the Expanding Universe. *MNRAS*, 108:252.

Bovy, J., Myers, A. D., Hennawi, J. F., Hogg, D. W., McMahon, R. G., Schiminovich, D., Sheldon, E. S., Brinkmann, J., Schneider, D. P., and Weaver, B. A. (2012). Photometric redshifts and quasar probabilities from a single, data-driven generative model. *The astrophysical journal*, 749(1):41.

Breiman, L. (2001). Random forests. *Machine learning*, 45(1):5–32.

Brooks, S. P. and Gelman, A. (1998). General methods for monitoring convergence of iterative simulations. *Journal of computational and graphical statistics*, 7(4):434–455.

Brown, A., Vallenari, A., Prusti, T., de Bruijne, J., Babusiaux, C., Bailer-Jones, C., Collaboration, G., et al. (2018). Gaia data release 2. summary of the contents and survey properties. *arXiv preprint arXiv:1804.09365*.

Castorina, E., Hand, N., Seljak, U., Beutler, F., Chuang, C.-H., Zhao, C., Gil-Marín, H., Percival, W. J., Ross, A. J., Choi, P. D., Dawson, K., de la Macorra, A., Rossi, G., Ruggeri, R., Schneider, D., and Zhao, G.-B. (2019). Redshift-weighted constraints on primordial non-Gaussianity from the clustering of the eBOSS DR14 quasars in Fourier space. *JCAP*, 2019(9):010.

Chabanier, S., Palanque-Delabrouille, N., Yèche, C., Le Goff, J.-M., Armengaud, E., Bautista, J., Blomqvist, M., Busca, N., Dawson, K., Etourneau, T., Font-Ribera, A., Lee, Y., du Mas des Bourboux, H., Pieri, M., Rich, J., Rossi, G., Schneider, D., and Slosar, A. (2019). The one-dimensional power spectrum from the SDSS DR14 Lyα forests. *JCAP*, 2019(7):017.

Chen, X., xin Huang, M., Kachru, S., and Shiu, G. (2007). Observational signatures and non-gaussianities of general single-field inflation. *Journal of Cosmology and Astroparticle Physics*, 2007(01):002–002.

Chevallier, M. and Polarski, D. (2001). Accelerating universes with scaling dark matter. *International Journal of Modern Physics D*, 10(02):213–223.

Chon, G., Challinor, A., Prunet, S., Hivon, E., and Szapudi, I. (2004). Fast estimation of polarization power spectra using correlation functions. *MNRAS*, 350(3):914–926.

Chuang, C.-H., Kitaura, F.-S., Prada, F., Zhao, C., and Yepes, G. (2015). Ezmocks: extending the zel'dovich approximation to generate mock galaxy catalogues with accurate clustering statistics. *Monthly Notices of the Royal Astronomical Society*, 446(3):2621–2628.

Clowe, D., Bradač, M., Gonzalez, A. H., Markevitch, M., Randall, S. W., Jones, C., and Zaritsky, D. (2006). A direct empirical proof of the existence of dark matter. *The Astrophysical Journal Letters*, 648(2):L109.

Cole, S., Percival, W. J., Peacock, J. A., Norberg, P., Baugh, C. M., Frenk, C. S., Baldry, I., Bland-Hawthorn, J., Bridges, T., Cannon, R., et al. (2005). The 2df galaxy redshift survey: power-spectrum analysis of the final data set and cosmological implications. *Monthly Notices of the Royal Astronomical Society*, 362(2):505–534.

Coles, P. and Jones, B. (1991). A lognormal model for the cosmological mass distribution. *MNRAS*, 248:1–13.

Colless, M., Dalton, G., Maddox, S., Sutherland, W., Norberg, P., Cole, S., Bland-Hawthorn, J., Bridges, T., Cannon, R., Collins, C., et al. (2001). The 2df galaxy redshift survey: spectra and redshifts. *Monthly Notices of the Royal Astronomical Society*, 328(4):1039–1063.

Comparat, J., Jullo, E., Kneib, J.-P., Schimd, C., Shan, H., Erben, T., Ilbert, O., Brownstein, J., Ealet, A., Escoffier, S., Moraes, B., Mostek, N., Newman, J. A., Pereira, M. E. S., Prada, F., Schlegel, D. J., Schneider, D. P., and Brandt, C. H. (2013). Stochastic bias of colour-selected BAO tracers by joint clustering-weak lensing analysis. *MNRAS*, 433(2):1146–1160.

Creminelli, P. and Zaldarriaga, M. (2004). A single-field consistency relation for the three-point function. *Journal of Cosmology and Astroparticle Physics*, 2004(10):006.

Crocce, M., Carretero, J., Bauer, A. H., Ross, A. J., Sevilla-Noarbe, I., Giannantonio, T., Sobreira, F., Sanchez, J., Gaztanaga, E., Carrasco Kind, M., Sánchez, C., Bonnett, C., Benoit-Lévy, A., Brunner, R. J., Carnero Rosell, A., Cawthon, R., Fosalba, P., Hartley, W., Kim, E. J., Leistedt, B., Miquel, R., Peiris, H. V., Percival, W. J., Rosenfeld, R., Rykoff, E. S., Sánchez, E., Abbott, T., Abdalla, F. B., Allam, S., Banerji, M., Bernstein, G. M., Bertin, E., Brooks, D., Buckley-Geer, E., Burke, D. L., Capozzi, D., Castand er, F. J., Cunha, C. E., D'Andrea, C. B., da Costa, L. N., Desai, S., Diehl, H. T., Eifler, T. F., Evrard, A. E., Fausti Neto, A., Fernand ez, E., Finley, D. A., Flaugher, B., Frieman, J., Gerdes, D. W., Gruen, D., Gruendl, R. A., Gutierrez, G., Honscheid, K., James, D. J., Kuehn, K., Kuropatkin, N., Lahav, O., Li, T. S., Lima, M., Maia, M. A. G., March, M., Marshall, J. L., Martini, P., Melchior, P., Miller, C. J., Neilsen, E., Nichol, R. C., Nord, B., Ogando, R., Plazas, A. A., Romer, A. K., Sako, M., Santiago, B., Schubnell, M., Smith, R. C., Soares-Santos, M., Suchyta, E., Swanson, M. E. C., Tarle, G., Thaler, J., Thomas, D., Vikram, V., Walker, A. R., Wechsler, R. H., Weller, J., Zuntz, J., and DES Collaboration (2016). Galaxy clustering, photometric redshifts and diagnosis of systematics in the DES Science Verification data. *MNRAS*, 455(4):4301–4324.

Cybenko, G. (1989). Approximation by superpositions of a sigmoidal function. *Mathematics of control, signals and systems*, 2(4):303–314.

Dahl, G. E., Sainath, T. N., and Hinton, G. E. (2013). Improving deep neural networks for lvcsr using rectified linear units and dropout. In *Acoustics, Speech and Signal Processing (ICASSP), 2013 IEEE International Conference on*, pages 8609–8613. IEEE.

Dalal, N., Dore, O., Huterer, D., and Shirokov, A. (2008). Imprints of primordial non-gaussianities on large-scale structure: Scale-dependent bias and abundance of virialized objects. *Physical Review D*, 77(12):123514.

Dawson, K. S., Kneib, J.-P., Percival, W. J., Alam, S., Albareti, F. D., Anderson, S. F., Armengaud, E., Aubourg, É., Bailey, S., Bautista, J. E., et al. (2016). The sdss-iv extended baryon oscillation spectroscopic survey: Overview and early data. *The Astronomical Journal*, 151(2):44.

Dawson, K. S., Schlegel, D. J., Ahn, C. P., Anderson, S. F., Aubourg, É., Bailey, S., Barkhouser, R. H., Bautista, J. E., Beifiori, A., Berlind, A. A., et al. (2012). The baryon oscillation spectroscopic survey of sdss-iii. *The Astronomical Journal*, 145(1):10.

de Sainte Agathe, V., Balland, C., du Mas des Bourboux, H., Blomqvist, M., Guy, J., Rich, J., Font-Ribera, A., Pieri, M. M., Bautista, J. E., Dawson, K., et al. (2019). Baryon acoustic oscillations at z= 2.34 from the correlations of ly α absorption in eboss dr14. *Astronomy and Astrophysics*, 629(BNL-212173-2019-JAAM).

Delubac, T., Raichoor, A., Comparat, J., Jouvel, S., Kneib, J.-P., Yèche, C., Zou, H., Brownstein, J., Abdalla, F., Dawson, K., et al. (2016). The sdss-iv eboss: Emission line galaxy catalogs at $z = 0.8$ and study of systematic errors in the angular clustering. *Monthly Notices of the Royal Astronomical Society*, page stw2741.

DESI Collaboration, Aghamousa, A., Aguilar, J., Ahlen, S., Alam, S., Allen, L. E., Allende Prieto, C., Annis, J., Bailey, S., Balland, C., Ballester, O., Baltay, C., Beaufore, L., Bebek, C., Beers, T. C., Bell, E. F., Bernal, J. L., Besuner, R., Beutler, F., Blake, C., Bleuler, H., Blomqvist, M., Blum, R., Bolton, A. S., Briceno, C., Brooks, D., Brownstein, J. R., Buckley-Geer, E., Burden, A., Burtin, E., Busca, N. G., Cahn, R. N., Cai, Y.-C., Cardiel-Sas, L., Carlberg, R. G., Carton, P.-H., Casas, R., Castander, F. J., Cervantes-Cota, J. L., Claybaugh, T. M., Close, M., Coker, C. T., Cole, S., Comparat, J., Cooper, A. P., Cousinou, M. C., Crocce, M., Cuby, J.-G., Cunningham, D. P., Davis, T. M., Dawson, K. S., de la Macorra, A., De Vicente, J., Delubac, T., Derwent, M., Dey, A., Dhungana, G., Ding, Z., Doel, P., Duan, Y. T., Ealet, A., Edelstein, J., Eftekharzadeh, S., Eisenstein, D. J., Elliott, A., Escoffier, S., Evatt, M., Fagrelius, P., Fan, X., Fanning, K., Farahi, A., Farihi, J., Favole, G., Feng, Y., Fernandez, E., Findlay, J. R., Finkbeiner, D. P., Fitzpatrick, M. J., Flaugher, B., Flender, S., Font-Ribera, A., Forero-Romero, J. E., Fosalba, P., Frenk, C. S., Fumagalli, M., Gaensicke, B. T., Gallo, G., Garcia Bellido, J., Gaztanaga, E., Pietro Gentile Fusillo, N., Gerard, T., Gershkovich, I., Giannantonio, T., Gillet, D., Gonzalez-de-Rivera, G., Gonzalez-Perez, V., Gott, S., Graur, O., Gutierrez, G., Guy, J., Habib, S., Heetderks, H., Heetderks, I., Heitmann, K., Hellwing, W. A., Herrera, D. A., Ho, S., Holland, S., Honscheid, K., Huff, E., Hutchinson, T. A., Huterer, D., Hwang, H. S., Illa Laguna, J. M., Ishikawa, Y., Jacobs, D., Jeffrey, N., Jelinsky, P., Jennings, E., Jiang, L., Jimenez, J., Johnson, J., Joyce, R., Jullo, E., Juneau, S., Kama, S., Karcher, A., Karkar, S., Kehoe, R., Kennamer, N., Kent, S., Kilbinger, M., Kim, A. G., Kirkby, D., Kisner, T., Kitanidis, E., Kneib, J.-P., Koposov, S., Kovacs, E., Koyama, K., Kremin, A., Kron, R., Kronig, L., Kueter-Young, A., Lacey, C. G., Lafever, R., Lahav, O., Lambert, A., Lampton, M., Landriau, M., Lang, D., Lauer, T. R., Le Goff, J.-M., Le Guillou, L., Le Van Suu, A., Lee, J. H., Lee, S.-J., Leitner, D., Lesser, M., Levi, M. E., L'Huillier, B., Li, B., Liang, M., Lin, H., Linder, E., Loebman, S. R., Lukić, Z., Ma, J., MacCrann, N., Magneville, C., Makarem, L., Manera, M., Manser, C. J., Marshall, R., Martini, P., Massey, R., Matheson, T., McCauley, J., McDonald, P., McGreer, I. D., Meisner, A., Metcalfe, N., Miller, T. N., Miquel, R., Moustakas, J., Myers, A., Naik, M., Newman, J. A., Nichol, R. C., Nicola, A., Nicolati da Costa, L., Nie, J., Niz, G., Norberg, P., Nord, B., Norman, D., Nugent, P., O'Brien, T., Oh, M., Olsen, K. A. G., Padilla, C., Padmanabhan, H., Padmanabhan, N., Palanque-Delabrouille, N., Palmese, A., Pappalardo, D., Pâris, I., Park, C., Patej, A., Peacock, J. A., Peiris, H. V., Peng, X., Percival, W. J., Perruchot, S., Pieri, M. M., Pogge, R., Pollack, J. E., Poppett, C., Prada, F., Prakash, A., Probst, R. G., Rabinowitz, D., Raichoor, A., Ree, C. H., Refregier, A., Regal, X., Reid, B., Reil, K., Rezaie, M., Rockosi, C. M., Roe, N., Ronayette, S., Roodman, A., Ross, A. J., Ross, N. P., Rossi, G., Rozo, E., Ruhlmann-Kleider, V., Rykoff, E. S., Sabiu, C., Samushia, L., Sanchez, E., Sanchez, J., Schlegel, D. J.,

Schneider, M., Schubnell, M., Secroun, A., Seljak, U., Seo, H.-J., Serrano, S., Shafieloo, A., Shan, H., Sharples, R., Sholl, M. J., Shourt, W. V., Silber, J. H., Silva, D. R., Sirk, M. M., Slosar, A., Smith, A., Smoot, G. F., Som, D., Song, Y.-S., Sprayberry, D., Staten, R., Stefanik, A., Tarle, G., Sien Tie, S., Tinker, J. L., Tojeiro, R., Valdes, F., Valenzuela, O., Valluri, M., Vargas-Magana, M., Verde, L., Walker, A. R., Wang, J., Wang, Y., Weaver, B. A., Weaverdyck, C., Wechsler, R. H., Weinberg, D. H., White, M., Yang, Q., Yeche, C., Zhang, T., Zhao, G.-B., Zheng, Y., Zhou, X., Zhou, Z., Zhu, Y., Zou, H., and Zu, Y. (2016). The DESI Experiment Part I: Science,Targeting, and Survey Design. *arXiv e-prints*, page arXiv:1611.00036.

Desjacques, V. and Seljak, U. (2010). Primordial non-gaussianity from the large-scale structure. *Classical and Quantum Gravity*, 27(12):124011.

Devijver, P. A. and Kittler, J. (1982). *Pattern recognition: A statistical approach*. Prentice hall.

Dey, A., Schlegel, D. J., Lang, D., Blum, R., Burleigh, K., Fan, X., Findlay, J. R., Finkbeiner, D., Herrera, D., Juneau, S., et al. (2018). Overview of the desi legacy imaging surveys. *arXiv preprint arXiv:1804.08657*.

Drinkwater, M. J., Jurek, R. J., Blake, C., Woods, D., Pimbblet, K. A., Glazebrook, K., Sharp, R., Pracy, M. B., Brough, S., Colless, M., et al. (2010). The wigglez dark energy survey: survey design and first data release. *Monthly Notices of the Royal Astronomical Society*, 401(3):1429–1452.

du Mas des Bourboux, H., Rich, J., Font-Ribera, A., de Sainte Agathe, V., Farr, J., Etourneau, T., Le Goff, J.-M., Cuceu, A., Balland, C., Bautista, J. E., Blomqvist, M., Brinkmann, J., Brownstein, J. R., Chabanier, S., Chaussidon, E., Dawson, K., González-Morales, A. X., Guy, J., Lyke, B. W., de la Macorra, A., Mueller, E.-M., Myers, A. D., Nitschelm, C., Muñoz Gutiérrez, A., Palanque-Delabrouille, N., Parker, J., Percival, W. J., Pérez-Ràfols, I., Petitjean, P., Pieri, M. M., Ravoux, C., Rossi, G., Schneider, D. P., Seo, H.-J., Slosar, A., Stermer, J., Vivek, M., Yèche, C., and Youles, S. (2020). The Completed SDSS-IV Extended Baryon Oscillation Spectroscopic Survey: Baryon Acoustic Oscillations with Lyα Forests. *ApJ*, 901(2):153.

Einstein, A. (1915). Die feldgleichungen der gravitation, preussische akademie der wissenschaften. *Sitzungsberichte*, page 844–847.

Einstein, A. (1917). Kosmologische betrachtungen zur allgemeinen relativitätstheorie. *Sitzungsberichte der Königlich Preußischen Akademie der Wissenschaften (Berlin), Seite 142-152.*

Eisenstein, D. J. (1997). An Analytic Expression for the Growth Function in a Flat Universe with a Cosmological Constant. *arXiv e-prints*, pages astro–ph/9709054.

Eisenstein, D. J. (2005). Dark energy and cosmic sound. *New Astronomy Reviews*, 49(7-9):360–365.

Eisenstein, D. J. and Hu, W. (1998). Baryonic features in the matter transfer function. *The Astrophysical Journal*, 496(2):605.

Eisenstein, D. J., Hu, W., and Tegmark, M. (1998). Cosmic complementarity: H0 and ωm from combining cosmic microwave background experiments and redshift surveys. *The Astrophysical Journal Letters*, 504(2):L57.

Eisenstein, D. J., Seo, H.-J., Sirko, E., and Spergel, D. N. (2007). Improving cosmological distance measurements by reconstruction of the baryon acoustic peak. *The Astrophysical Journal*, 664(2):675.

Eisenstein, D. J., Seo, H.-J., and White, M. (2007). On the Robustness of the Acoustic Scale in the Low-Redshift Clustering of Matter. *ApJ*, 664(2):660–674.

Eisenstein, D. J., Zehavi, I., Hogg, D. W., Scoccimarro, R., Blanton, M. R., Nichol, R. C., Scranton, R., Seo, H.-J., Tegmark, M., Zheng, Z., et al. (2005). Detection of the baryon acoustic peak in the large-scale correlation function of sdss luminous red galaxies *The Astrophysical Journal*, 633(2):560.

Elsner, F., Leistedt, B., and Peiris, H. V. (2015). Unbiased methods for removing systematics from galaxy clustering measurements. *Monthly Notices of the Royal Astronomical Society*, 456(2):2095–2104.

Elvin-Poole, J., Crocce, M., Ross, A. J., Giannantonio, T., Rozo, E., Rykoff, E. S., Avila, S., Banik, N., Blazek, J., Bridle, S. L., Cawthon, R., Drlica-Wagner, A., Friedrich, O., Kokron, N., Krause, E., MacCrann, N., Prat, J., Sánchez, C., Secco, L. F., Sevilla-Noarbe, I., Troxel, M. A., Abbott, T. M. C., Abdalla, F. B., Allam, S., Annis, J., Asorey, J., Bechtol, K., Becker, M. R., Benoit-Lévy, A., Bernstein, G. M., Bertin, E., Brooks, D., Buckley-Geer, E., Burke, D. L., Carnero Rosell, A., Carollo, D., Carrasco Kind, M., Carretero, J., Castander, F. J., Cunha, C. E., D'Andrea, C. B., da Costa, L. N., Davis, T. M., Davis, C., Desai, S., Diehl, H. T., Dietrich, J. P., Dodelson, S., Doel, P., Eifler, T. F., Evrard, A. E., Fernandez, E., Flaugher, B., Fosalba, P., Frieman, J., García-Bellido, J., Gaztanaga, E., Gerdes, D. W., Glazebrook, K., Gruen, D., Gruendl, R. A., Gschwend, J., Gutierrez, G., Hartley, W. G., Hinton, S. R., Honscheid, K., Hoormann, J. K., Jain, B., James, D. J., Jarvis, M., Jeltema, T., Johnson, M. W. G., Johnson, M. D., King, A., Kuehn, K., Kuhlmann, S., Kuropatkin, N., Lahav, O., Lewis, G., Li, T. S., Lidman, C., Lima, M., Lin, H., Macaulay, E., March, M., Marshall, J. L., Martini, P., Melchior, P., Menanteau, F., Miquel, R., Mohr, J. J., Möller, A., Nichol, R. C., Nord, B., O'Neill, C. R., Percival, W. J., Petravick, D., Plazas, A. A., Romer, A. K., Sako, M., Sanchez, E., Scarpine, V., Schindler, R., Schubnell, M., Sheldon, E., Smith, M., Smith, R. C., Soares-Santos, M., Sobreira, F., Sommer, N. E., Suchyta, E., Swanson, M. E. C., Tarle, G., Thomas, D., Tucker, B. E., Tucker, D. L., Uddin, S. A., Vikram, V., Walker, A. R.,

Wechsler, R. H., Weller, J., Wester, W., Wolf, R. C., Yuan, F., Zhang, B., Zuntz, J., and DES Collaboration (2018a). Dark Energy Survey year 1 results: Galaxy clustering for combined probes. *PRD*, 98(4):042006.

Elvin-Poole, J., Crocce, M., Ross, A. J., Giannantonio, T., Rozo, E., Rykoff, E. S., Avila, S., Banik, N., Blazek, J., Bridle, S. L., Cawthon, R., Drlica-Wagner, A., Friedrich, O., Kokron, N., Krause, E., MacCrann, N., Prat, J., Sánchez, C., Secco, L. F., Sevilla-Noarbe, I., Troxel, M. A., Abbott, T. M. C., Abdalla, F. B., Allam, S., Annis, J., Asorey, J., Bechtol, K., Becker, M. R., Benoit-Lévy, A., Bernstein, G. M., Bertin, E., Brooks, D., Buckley-Geer, E., Burke, D. L., Carnero Rosell, A., Carollo, D., Carrasco Kind, M., Carretero, J., Castander, F. J., Cunha, C. E., D'Andrea, C. B., da Costa, L. N., Davis, T. M., Davis, C., Desai, S., Diehl, H. T., Dietrich, J. P., Dodelson, S., Doel, P., Eifler, T. F., Evrard, A. E., Fernandez, E., Flaugher, B., Fosalba, P., Frieman, J., García-Bellido, J., Gaztanaga, E., Gerdes, D. W., Glazebrook, K., Gruen, D., Gruendl, R. A., Gschwend, J., Gutierrez, G., Hartley, W. G., Hinton, S. R., Honscheid, K., Hoormann, J. K., Jain, B., James, D. J., Jarvis, M., Jeltema, T., Johnson, M. W. G., Johnson, M. D., King, A., Kuehn, K., Kuhlmann, S., Kuropatkin, N., Lahav, O., Lewis, G., Li, T. S., Lidman, C., Lima, M., Lin, H., Macaulay, E., March, M., Marshall, J. L., Martini, P., Melchior, P., Menanteau, F., Miquel, R., Mohr, J. J., Möller, A., Nichol, R. C., Nord, B., O'Neill, C. R., Percival, W. J., Petravick, D., Plazas, A. A., Romer, A. K., Sako, M., Sanchez, E., Scarpine, V., Schindler, R., Schubnell, M., Sheldon, E., Smith, M., Smith, R. C., Soares-Santos, M., Sobreira, F., Sommer, N. E., Suchyta, E., Swanson, M. E. C., Tarle, G., Thomas, D., Tucker, B. E., Tucker, D. L., Uddin, S. A., Vikram, V., Walker, A. R., Wechsler, R. H., Weller, J., Wester, W., Wolf, R. C., Yuan, F., Zhang, B., Zuntz, J., and DES Collaboration (2018b). Dark Energy Survey year 1 results: Galaxy clustering for combined probes. *PRD*, 98(4):042006.

Falk, T., Rangarajan, R., and Srednicki, M. (1992). Dependence of density perturbations on the coupling constant in a simple model of inflation. *Physical Review D*, 46(10):4232.

Fang, X., Krause, E., Eifler, T., and MacCrann, N. (2020). Beyond Limber: efficient computation of angular power spectra for galaxy clustering and weak lensing. *JCAP*, 2020(5):010.

Feldman, H. A., Kaiser, N., and Peacock, J. A. (1994). Power-Spectrum Analysis of Three-dimensional Redshift Surveys. *ApJ*, 426:23.

Fixsen, D. J., Cheng, E. S., Cottingham, D. A., Eplee, Jr., R. E., Isaacman, R. B., Mather, J. C., Meyer, S. S., Noerdlinger, P. D., Shafer, R. A., Weiss, R., Wright, E. L., Bennett, C. L., Boggess, N. W., Kelsall, T., Moseley, S. H., Silverberg, R. F., Smoot, G. F., and Wilkinson, D. T. (1994). Cosmic microwave background dipole spectrum measured by the COBE FIRAS instrument. *APJ*, 420:445–449.

Flaugher, B. (2005). The Dark Energy Survey. *International Journal of Modern Physics A*, 20(14):3121–3123.

Friedman, A. (1922). Über die krümmung des raumes. *Zeitschrift für Physik A Hadrons and Nuclei*, 10(1):377–386.

Friedmann, A. (1924). Über die möglichkeit einer welt mit konstanter negativer krümmung des raumes. *Zeitschrift für Physik A Hadrons and Nuclei*, 21(1):326–332.

Fukuda, Y., Hayakawa, T., Ichihara, E., Inoue, K., Ishihara, K., Ishino, H., Itow, Y., Kajita, T., Kameda, J., Kasuga, S., et al. (1998). Evidence for oscillation of atmospheric neutrinos. *Physical Review Letters*, 81(8):1562.

Fukugita, M., Shimasaku, K., Ichikawa, T., Gunn, J., et al. (1996). The sloan digital sky survey photometric system. Technical report, SCAN-9601313.

Funahashi, K.-I. (1989). On the approximate realization of continuous mappings by neural networks. *Neural networks*, 2(3):183–192.

Gaia Collaboration, Brown, A. G. A., Vallenari, A., Prusti, T., de Bruijne, J. H. J., Babusiaux, C., Bailer-Jones, C. A. L., Biermann, M., Evans, D. W., Eyer, L., Jansen, F., Jordi, C., Klioner, S. A., Lammers, U., Lindegren, L., Luri, X., Mignard, F., Panem, C., Pourbaix, D., Randich, S., Sartoretti, P., Siddiqui, H. I., Soubiran, C., van Leeuwen, F., Walton, N. A., Arenou, F., Bastian, U., Cropper, M., Drimmel, R., Katz, D., Lattanzi, M. G., Bakker, J., Cacciari, C., Castañeda, J., Chaoul, L., Cheek, N., De Angeli, F., Fabricius, C., Guerra, R., Holl, B., Masana, E., Messineo, R., Mowlavi, N., Nienartowicz, K., Panuzzo, P., Portell, J., Riello, M., Seabroke, G. M., Tanga, P., Thévenin, F., Gracia-Abril, G., Comoretto, G., Garcia-Reinaldos, M., Teyssier, D., Altmann, M., Andrae, R., Audard, M., Bellas-Velidis, I., Benson, K., Berthier, J., Blomme, R., Burgess, P., Busso, G., Carry, B., Cellino, A., Clementini, G., Clotet, M., Creevey, O., Davidson, M., De Ridder, J., Delchambre, L., Dell'Oro, A., Ducourant, C., Fernández-Hernández, J., Fouesneau, M., Frémat, Y., Galluccio, L., García-Torres, M., González-Núñez, J., González-Vidal, J. J., Gosset, E., Guy, L. P., Halbwachs, J. L., Hambly, N. C., Harrison, D. L., Hernández, J., Hestroffer, D., Hodgkin, S. T., Hutton, A., Jasniewicz, G., Jean-Antoine-Piccolo, A., Jordan, S., Korn, A. J., Krone-Martins, A., Lanzafame, A. C., Lebzelter, T., Löffler, W., Manteiga, M., Marrese, P. M., Martín-Fleitas, J. M., Moitinho, A., Mora, A., Muinonen, K., Osinde, J., Pancino, E., Pauwels, T., Petit, J. M., Recio-Blanco, A., Richards, P. J., Rimoldini, L., Robin, A. C., Sarro, L. M., Siopis, C., Smith, M., Sozzetti, A., Süveges, M., Torra, J., van Reeven, W., Abbas, U., Abreu Aramburu, A., Accart, S., Aerts, C., Altavilla, G., Álvarez, M. A., Alvarez, R., Alves, J., Anderson, R. I., Andrei, A. H., Anglada Varela, E., Antiche, E., Antoja, T., Arcay, B., Astraatmadja, T. L., Bach, N., Baker, S. G., Balaguer-Núñez, L., Balm, P., Barache, C., Barata, C., Barbato, D., Barblan, F., Barklem, P. S., Barrado, D., Barros, M., Barstow, M. A., Bartholomé Muñoz, S., Bassilana, J. L., Becciani, U., Bellazzini, M., Berihuete, A., Bertone, S., Bianchi, L., Bienaymé, O., Blanco-Cuaresma, S., Boch, T., Boeche, C., Bombrun, A., Borrachero, R., Bossini, D., Bouquillon, S., Bourda, G., Bragaglia, A., Bramante, L., Breddels, M. A., Bressan, A., Brouillet, N., Brüsemeister,

T., Brugaletta, E., Bucciarelli, B., Burlacu, A., Busonero, D., Butkevich, A. G., Buzzi, R., Caffau, E., Cancelliere, R., Cannizzaro, G., Cantat-Gaudin, T., Carballo, R., Carlucci, T., Carrasco, J. M., Casamiquela, L., Castellani, M., Castro-Ginard, A., Charlot, P., Chemin, L., Chiavassa, A., Cocozza, G., Costigan, G., Cowell, S., Crifo, F., Crosta, M., Crowley, C., Cuypers, J., Dafonte, C., Damerdji, Y., Dapergolas, A., David, P., David, M., de Laverny, P., De Luise, F., De March, R., de Martino, D., de Souza, R., de Torres, A., Debosscher, J., del Pozo, E., Delbo, M., Delgado, A., Delgado, H. E., Di Matteo, P., Diakite, S., Diener, C., Distefano, E., Dolding, C., Drazinos, P., Durán, J., Edvardsson, B., Enke, H., Eriksson, K., Esquej, P., Eynard Bontemps, G., Fabre, C., Fabrizio, M., Faigler, S., Falcão, A. J., Farràs Casas, M., Federici, L., Fedorets, G., Fernique, P., Figueras, F., Filippi, F., Findeisen, K., Fonti, A., Fraile, E., Fraser, M., Frézouls, B., Gai, M., Galleti, S., Garabato, D., García-Sedano, F., Garofalo, A., Garralda, N., Gavel, A., Gavras, P., Gerssen, J., Geyer, R., Giacobbe, P., Gilmore, G., Girona, S., Giuffrida, G., Glass, F., Gomes, M., Granvik, M., Gueguen, A., Guerrier, A., Guiraud, J., Gutiérrez-Sánchez, R., Haigron, R., Hatzidimitriou, D., Hauser, M., Haywood, M., Heiter, U., Helmi, A., Heu, J., Hilger, T., Hobbs, D., Hofmann, W., Holland, G., Huckle, H. E., Hypki, A., Icardi, V., Janßen, K., Jevardat de Fombelle, G., Jonker, P. G., Juhász, Á. L., Julbe, F., Karampelas, A., Kewley, A., Klar, J., Kochoska, A., Kohley, R., Kolenberg, K., Kontizas, M., Kontizas, E., Koposov, S. E., Kordopatis, G., Kostrzewa-Rutkowska, Z., Koubsky, P., Lambert, S., Lanza, A. F., Lasne, Y., Lavigne, J. B., Le Fustec, Y., Le Poncin-Lafitte, C., Lebreton, Y., Leccia, S., Leclerc, N., Lecoeur-Taibi, I., Lenhardt, H., Leroux, F., Liao, S., Licata, E., Lindstrøm, H. E. P., Lister, T. A., Livanou, E., Lobel, A., López, M., Managau, S., Mann, R. G., Mantelet, G., Marchal, O., Marchant, J. M., Marconi, M., Marinoni, S., Marschalkó, G., Marshall, D. J., Martino, M., Marton, G., Mary, N., Massari, D., Matijevič, G., Mazeh, T., McMillan, P. J., Messina, S., Michalik, D., Millar, N. R., Molina, D., Molinaro, R., Molnár, L., Montegriffo, P., Mor, R., Morbidelli, R., Morel, T., Morris, D., Mulone, A. F., Muraveva, T., Musella, I., Nelemans, G., Nicastro, L., Noval, L., O'Mullane, W., Ordénovic, C., Ordóñez-Blanco, D., Osborne, P., Pagani, C., Pagano, I., Pailler, F., Palacin, H., Palaversa, L., Panahi, A., Pawlak, M., Piersimoni, A. M., Pineau, F. X., Plachy, E., Plum, G., Poggio, E., Poujoulet, E., Prša, A., Pulone, L., Racero, E., Ragaini, S., Rambaux, N., Ramos-Lerate, M., Regibo, S., Reylé, C., Riclet, F., Ripepi, V., Riva, A., Rivard, A., Rixon, G., Roegiers, T., Roelens, M., Romero-Gómez, M., Rowell, N., Royer, F., Ruiz-Dern, L., Sadowski, G., Sagristà Sellés, T., Sahlmann, J., Salgado, J., Salguero, E., Sanna, N., Santana-Ros, T., Sarasso, M., Savietto, H., Schultheis, M., Sciacca, E., Segol, M., Segovia, J. C., Ségransan, D., Shih, I. C., Siltala, L., Silva, A. F., Smart, R. L., Smith, K. W., Solano, E., Solitro, F., Sordo, R., Soria Nieto, S., Souchay, J., Spagna, A., Spoto, F., Stampa, U., Steele, I. A., Steidelmüller, H., Stephenson, C. A., Stoev, H., Suess, F. F., Surdej, J., Szabados, L., Szegedi-Elek, E., Tapiador, D., Taris, F., Tauran, G., Taylor, M. B., Teixeira, R., Terrett, D., Teyssand ier, P., Thuillot, W., Titarenko, A., Torra Clotet, F., Turon, C., Ulla, A., Utrilla, E., Uzzi, S., Vaillant, M., Valentini, G., Valette, V., van Elteren, A., Van Hemelryck, E., van Leeuwen, M., Vaschetto, M., Vecchiato, A., Veljanoski, J., Viala, Y., Vicente, D., Vogt, S., von Essen, C., Voss, H., Votruba,

V., Voutsinas, S., Walmsley, G., Weiler, M., Wertz, O., Wevers, T., Wyrzykowski, Ł., Yoldas, A., Žerjal, M., Ziaeepour, H., Zorec, J., Zschocke, S., Zucker, S., Zurbach, C., and Zwitter, T. (2018). Gaia Data Release 2. Summary of the contents and survey properties. *AAP*, 616:A1.

Gangui, A., Lucchin, F., Matarrese, S., and Mollerach, S. (1993). The three–point correlation function of the cosmic microwave background in inflationary models. *arXiv preprint astro-ph/9312033*.

Geach, J. E., Smail, I., Best, P. N., Kurk, J., Casali, M., Ivison, R. J., and Coppin, K. (2008). HiZELS: a high-redshift survey of Hα emitters - I. The cosmic star formation rate and clustering at z = 2.23. *MNRAS*, 388(4):1473–1486.

Gelman, A., Rubin, D. B., et al. (1992). Inference from iterative simulation using multiple sequences. *Statistical science*, 7(4):457–472.

Geurts, P., Ernst, D., and Wehenkel, L. (2006). Extremely randomized trees. *Machine learning*, 63(1):3–42.

Giannantonio, T., Ross, A. J., Percival, W. J., Crittenden, R., Bacher, D., Kilbinger, M., Nichol, R., and Weller, J. (2014). Improved primordial non-gaussianity constraints from measurements of galaxy clustering and the integrated sachs-wolfe effect. *Physical Review D*, 89(2):023511.

Giunti, C. and Kim, C. W. (2007). *Fundamentals of neutrino physics and astrophysics*. Oxford university press.

Glorot, X., Bordes, A., and Bengio, Y. (2011). Deep sparse rectifier neural networks. In *Proceedings of the fourteenth international conference on artificial intelligence and statistics*, pages 315–323.

Gorski, K. M., Hivon, E., Banday, A., Wandelt, B. D., Hansen, F. K., Reinecke, M., and Bartelmann, M. (2005). Healpix: a framework for high-resolution discretization and fast analysis of data distributed on the sphere. *The Astrophysical Journal*, 622(2):759.

Grossi, M., Branchini, E., Dolag, K., Matarrese, S., and Moscardini, L. (2008). The mass density field in simulated non-gaussian scenarios. *Monthly Notices of the Royal Astronomical Society*, 390(1):438–446.

Gunn, J. E., Siegmund, W. A., Mannery, E. J., Owen, R. E., Hull, C. L., Leger, R. F., Carey, L. N., Knapp, G. R., York, D. G., Boroski, W. N., et al. (2006). The 2.5 m telescope of the sloan digital sky survey. *The Astronomical Journal*, 131(4):2332.

Guth, A. H. and Pi, S.-Y. (1982). Fluctuations in the new inflationary universe. *Physical Review Letters*, 49(15):1110.

Guyon, I. and Elisseeff, A. (2003). An introduction to variable and feature selection. *Journal of machine learning research*, 3(Mar):1157–1182.

Hahn, C., Scoccimarro, R., Blanton, M. R., Tinker, J. L., and Rodríguez-Torres, S. A. (2017). The effect of fibre collisions on the galaxy power spectrum multipoles. *Monthly Notices of the Royal Astronomical Society*, 467(2):1940–1956.

Hamilton, A. (1998). Linear redshift distortions: a review. In *The evolving universe*, pages 185–275. Springer.

Hamilton, A. J. S. (2005). Power Spectrum Estimation I. Basics. *ArXiv Astrophysics e-prints*.

Hand, N., Feng, Y., Beutler, F., Li, Y., Modi, C., Seljak, U., and Slepian, Z. (2017). nbodykit: an open-source, massively parallel toolkit for large-scale structure. *arXiv preprint arXiv:1712.05834*.

Hand, N., Li, Y., Slepian, Z., and Seljak, U. (2017). An optimal FFT-based anisotropic power spectrum estimator. *JCAP*, 2017(7):002.

Hartlap, J., Simon, P., and Schneider, P. (2007). Why your model parameter confidences might be too optimistic. Unbiased estimation of the inverse covariance matrix. *AAP*, 464(1):399–404.

Hawking, S. (1982). The development of irregularities in a single bubble inflationary universe.

He, K., Zhang, X., Ren, S., and Sun, J. (2015). Delving deep into rectifiers: Surpassing human-level performance on imagenet classification. In *Proceedings of the IEEE international conference on computer vision*, pages 1026–1034.

HI4PI Collaboration, Ben Bekhti, N., Flöer, L., Keller, R., Kerp, J., Lenz, D., Winkel, B., Bailin, J., Calabretta, M. R., Dedes, L., Ford, H. A., Gibson, B. K., Haud, U., Janowiecki, S., Kalberla, P. M. W., Lockman, F. J., McClure-Griffiths, N. M., Murphy, T., Nakanishi, H., Pisano, D. J., and Staveley-Smith, L. (2016). HI4PI: A full-sky H I survey based on EBHIS and GASS. *AAP*, 594:A116.

Hivon, E., Górski, K. M., Netterfield, C. B., Crill, B. P., Prunet, S., and Hansen, F. (2002). MASTER of the Cosmic Microwave Background Anisotropy Power Spectrum: A Fast Method for Statistical Analysis of Large and Complex Cosmic Microwave Background Data Sets. *ApJ*, 567(1):2–17.

Ho, S., Agarwal, N., Myers, A. D., Lyons, R., Disbrow, A., Seo, H.-J., Ross, A., Hirata, C., Padmanabhan, N., O'Connell, R., Huff, E., Schlegel, D., Slosar, A., Weinberg, D., Strauss, M., Ross, N. P., Schneider, D. P., Bahcall, N., Brinkmann, J., Palanque-Delabrouille, N., and Yèche, C. (2015). Sloan digital sky survey III photometric quasar clustering: probing the initial conditions of the universe. *Journal of Cosmology and Astroparticle Physics*, 2015(05):040–040.

Ho, S., Cuesta, A., Seo, H.-J., de Putter, R., Ross, A. J., White, M., Padmanabhan, N., Saito, S., Schlegel, D. J., Schlafly, E., Seljak, U., Hernández-Monteagudo, C., Sánchez, A. G., Percival, W. J., Blanton, M., Skibba, R., Schneider, D., Reid, B., Mena, O., Viel, M., Eisenstein, D. J., Prada, F., Weaver, B. A., Bahcall, N., Bizyaev, D., Brewinton, H., Brinkman, J., Nicolaci da Costa, L., Gott, J. R., Malanushenko, E., Malanushenko, V., Nichol, B., Oravetz, D., Pan, K., Palanque-Delabrouille, N., Ross, N. P., Simmons, A., de Simoni, F., Snedden, S., and Yeche, C. (2012). Clustering of Sloan Digital Sky Survey III Photometric Luminous Galaxies: The Measurement, Systematics, and Cosmological Implications. *APJ*, 761:14.

Ho, S., Hirata, C., Padmanabhan, N., Seljak, U., and Bahcall, N. (2008). Correlation of cmb with large-scale structure. i. integrated sachs-wolfe tomography and cosmological implications. *Physical Review D*, 78(4):043519.

Hobson, E. W. (1931). *The theory of spherical and ellipsoidal harmonics*. CUP Archive.

Hoerl, A. E. and Kennard, R. W. (1970). Ridge regression: Biased estimation for nonorthogonal problems. *Technometrics*, 12(1):55–67.

Hornik, K., Stinchcombe, M., and White, H. (1989). Multilayer feedforward networks are universal approximators. *Neural networks*, 2(5):359–366.

Hou, J., Sánchez, A. G., Ross, A. J., Smith, A., Neveux, R., Bautista, J., Burtin, E., Zhao, C., Scoccimarro, R., Dawson, K. S., de Mattia, A., de la Macorra, A., du Mas des Bourboux, H., Eisenstein, D. J., Gil-Marín, H., Lyke, B. W., Mohammad, F. G., Mueller, E.-M., Percival, W. J., Rossi, G., Vargas Magaña, M., Zarrouk, P., Zhao, G.-B., Brinkmann, J., Brownstein, J. R., Chuang, C.-H., Myers, A. D., Newman, J. A., Schneider, D. P., and Vivek, M. (2021). The completed SDSS-IV extended Baryon Oscillation Spectroscopic Survey: BAO and RSD measurements from anisotropic clustering analysis of the quasar sample in configuration space between redshift 0.8 and 2.2. *MNRAS*, 500(1):1201–1221.

Hoyle, F. (1948). A New Model for the Expanding Universe. *MNRAS*, 108:372.

Huang, G.-B. (2003). Learning capability and storage capacity of two-hidden-layer feedforward networks. *IEEE Transactions on Neural Networks*, 14(2):274–281.

Hubble, E. (1926). No. 324. extra-galactic nebulae. *Contributions from the Mount Wilson Observatory/Carnegie Institution of Washington*, 324:1–49.

Hubble, E. (1929). A relation between distance and radial velocity among extra-galactic nebulae. *Proceedings of the National Academy of Sciences*, 15(3):168–173.

Hubble, E. and Humason, M. L. (1931). The velocity-distance relation among extra-galactic nebulae. *The Astrophysical Journal*, 74:43.

Hutchinson, T. A., Bolton, A. S., Dawson, K. S., Prieto, C. A., Bailey, S., Bautista, J. E., Brownstein, J. R., Conroy, C., Guy, J., Myers, A. D., et al. (2016). Redshift measurement and spectral classification for eboss galaxies with the redmonster software. *The Astronomical Journal*, 152(6):205.

Huterer, D., Cunha, C. E., and Fang, W. (2013). Calibration errors unleashed: effects on cosmological parameters and requirements for large-scale structure surveys. *Monthly Notices of the Royal Astronomical Society*, 432(4):2945–2961.

Ioffe, S. and Szegedy, C. (2015). Batch Normalization: Accelerating Deep Network Training by Reducing Internal Covariate Shift. *arXiv e-prints*, page arXiv:1502.03167.

Ivezic, Z., Tyson, J., Abel, B., Acosta, E., Allsman, R., AlSayyad, Y., Anderson, S., Andrew, J., Angel, R., Angeli, G., et al. (2008). Lsst: from science drivers to reference design and anticipated data products. *arXiv preprint arXiv:0805.2366*.

Jensen, T. W., Vivek, M., Dawson, K. S., Anderson, S. F., Bautista, J., Bizyaev, D., Brandt, W. N., Brownstein, J. R., Green, P., Harris, D. W., et al. (2016). Spectral evolution in high redshift quasars from the final baryon oscillation spectroscopic survey sample. *The Astrophysical Journal*, 833(2):199.

Jing, Y. P. (2005). Correcting for the Alias Effect When Measuring the Power Spectrum Using a Fast Fourier Transform. *ApJ*, 620(2):559–563.

John, G. H., Kohavi, R., and Pfleger, K. (1994). Irrelevant features and the subset selection problem. In *Machine Learning Proceedings 1994*, pages 121–129. Elsevier.

Kaiser, N. (1984). On the spatial correlations of Abell clusters. *ApJl*, 284:L9–L12.

Kaiser, N. (1987). Clustering in real space and in redshift space. *Monthly Notices of the Royal Astronomical Society*, 227(1):1–21.

Kalus, B., Percival, W., Bacon, D., Mueller, E., Samushia, L., Verde, L., Ross, A., and Bernal, J. (2018). A map-based method for eliminating systematic modes from galaxy clustering power spectra with application to boss. *Monthly Notices of the Royal Astronomical Society*, 482(1):453–470.

Kalus, B., Percival, W., Bacon, D., Mueller, E., Samushia, L., Verde, L., Ross, A., and Bernal, J. (2019). A map-based method for eliminating systematic modes from galaxy clustering power spectra with application to boss. *Monthly Notices of the Royal Astronomical Society*, 482(1):453–470.

Kalus, B., Percival, W. J., Bacon, D., and Samushia, L. (2016). Unbiased contaminant removal for 3d galaxy power spectrum measurements. *Monthly Notices of the Royal Astronomical Society*, 463(1):467–476.

Kalus, B., Percival, W. J., Bacon, D. J., and Samushia, L. (2016). Unbiased contaminant removal for 3D galaxy power spectrum measurements. *MNRAS*, 463(1):467–476.

Karagiannis, D., Shanks, T., and Ross, N. P. (2014). Search for primordial non-gaussianity in the quasars of sdss-iii boss dr9. *Monthly Notices of the Royal Astronomical Society*, 441(1):486–502.

Kerscher, M., Szapudi, I., and Szalay, A. S. (2000). A Comparison of Estimators for the Two-Point Correlation Function. *ApJ*, 535:L13–L16.

Kingma, D. P. and Ba, J. (2014). Adam: A method for stochastic optimization. *arXiv preprint arXiv:1412.6980*.

Kofman, L. and Pogosyan, D. Y. (1988). Nonflat perturbations in inflationary cosmology. *Physics Letters B*, 214(4):508–514.

Kohavi, R. and John, G. H. (1997). Wrappers for feature subset selection. *Artificial intelligence*, 97(1-2):273–324.

Koller, D. and Sahami, M. (1996). Toward optimal feature selection. Technical report, Stanford InfoLab.

Komatsu, E. and Spergel, D. N. (2001). Acoustic signatures in the primary microwave background bispectrum. *Physical Review D*, 63(6):063002.

Krizhevsky, A., Sutskever, I., and Hinton, G. E. (2012). Imagenet classification with deep convolutional neural networks. In *Advances in neural information processing systems*, pages 1097–1105.

Landy, S. D. and Szalay, A. S. (1993). Bias and variance of angular correlation functions. *The Astrophysical Journal*, 412:64–71.

Lang, D., Hogg, D. W., and Mykytyn, D. (2016). The tractor: Probabilistic astronomical source detection and measurement. *Astrophysics Source Code Library*.

Laurent, P., Eftekharzadeh, S., Le Goff, J.-M., Myers, A., Burtin, E., White, M., Ross, A. J., Tinker, J., Tojeiro, R., Bautista, J., et al. (2017). Clustering of quasars in sdss-iv eboss: study of potential systematics and bias determination. *Journal of Cosmology and Astroparticle Physics*, 2017(07):017.

Leavitt, H. S. (1908). 1777 variables in the Magellanic Clouds. *Annals of Harvard College Observatory*, 60:87–108.3.

Leavitt, H. S. and Pickering, E. C. (1912). Periods of 25 Variable Stars in the Small Magellanic Cloud. *Harvard College Observatory Circular*, 173:1–3.

Leistedt, B. and Peiris, H. V. (2014). Exploiting the full potential of photometric quasar surveys: optimal power spectra through blind mitigation of systematics. *Monthly Notices of the Royal Astronomical Society*, 444(1):2–14.

Leistedt, B., Peiris, H. V., Elsner, F., Benoit-Lévy, A., Amara, A., Bauer, A. H., Becker, M. R., Bonnett, C., Bruderer, C., Busha, M. T., Carrasco Kind, M., Chang, C., Crocce, M., da Costa, L. N., Gaztanaga, E., Huff, E. M., Lahav, O., Palmese, A., Percival, W. J., Refregier, A., Ross, A. J., Rozo, E., Rykoff, E. S., Sánchez, C., Sadeh, I., Sevilla-Noarbe, I., Sobreira, F., Suchyta, E., Swanson, M. E. C., Wechsler, R. H., Abdalla, F. B., Allam, S., Banerji, M., Bernstein, G. M., Bernstein, R. A., Bertin, E., Bridle, S. L., Brooks, D., Buckley-Geer, E., Burke, D. L., Capozzi, D., Carnero Rosell, A., Carretero, J., Cunha, C. E., D'Andrea, C. B., DePoy, D. L., Desai, S., Diehl, H. T., Doel, P., Eifler, T. F., Evrard, A. E., Fausti Neto, A., Flaugher, B., Fosalba, P., Frieman, J., Gerdes, D. W., Gruen, D., Gruendl, R. A., Gutierrez, G., Honscheid, K., James, D. J., Jarvis, M., Kent, S., Kuehn, K., Kuropatkin, N., Li, T. S., Lima, M., Maia, M. A. G., March, M., Marshall, J. L., Martini, P., Melchior, P., Miller, C. J., Miquel, R., Nichol, R. C., Nord, B., Ogando, R., Plazas, A. A., Reil, K., Romer, A. K., Roodman, A., Sanchez, E., Santiago, B., Scarpine, V., Schubnell, M., Smith, R. C., Soares-Santos, M., Tarle, G., Thaler, J., Thomas, D., Vikram, V., Walker, A. R., Wester, W., Zhang, Y., and Zuntz, J. (2016). Mapping and Simulating Systematics due to Spatially Varying Observing Conditions in DES Science Verification Data. *ApJs*, 226:24.

Leistedt, B., Peiris, H. V., Mortlock, D. J., Benoit-Lévy, A., and Pontzen, A. (2013). Estimating the large-scale angular power spectrum in the presence of systematics: a case study of sloan digital sky survey quasars. *Monthly Notices of the Royal Astronomical Society*, 435(3):1857–1873.

Leistedt, B., Peiris, H. V., and Roth, N. (2014). Constraints on primordial non-gaussianity from 800 000 photometric quasars. *Physical Review Letters*, 113(22):221301.

Lemaître, G. (1927). Un univers homogène de masse constante et de rayon croissant rendant compte de la vitesse radiale des nébuleuses extra-galactiques. In *Annales de la Société scientifique de Bruxelles*, volume 47, pages 49–59.

Lesgourgues, J. and Pastor, S. (2006). Massive neutrinos and cosmology. *Physics Reports*, 429(6):307–379.

Lin, H. W., Tegmark, M., and Rolnick, D. (2017). Why does deep and cheap learning work so well? *Journal of Statistical Physics*, 168(6):1223–1247.

Linde, A. and Mukhanov, V. (1997). Non-gaussian isocurvature perturbations from inflation. *Physical Review D*, 56(2):R535.

Linder, E. V. (2003). Exploring the expansion history of the universe. *Physical Review Letters*, 90(9):091301.

Loshchilov, I. and Hutter, F. (2016). SGDR: Stochastic Gradient Descent with Warm Restarts. *arXiv e-prints*, page arXiv:1608.03983.

Loshchilov, I. and Hutter, F. (2017). Decoupled Weight Decay Regularization. *arXiv e-prints*, page arXiv:1711.05101.

LSST Science Collaborations, Marshall, P., Anguita, T., Bianco, F. B., Bellm, E. C., Brandt, N., Clarkson, W., Connolly, A., Gawiser, E., Ivezic, Z., Jones, L., Lochner, M., Lund, M. B., Mahabal, A., Nidever, D., Olsen, K., Ridgway, S., Rhodes, J., Shemmer, O., Trilling, D., Vivas, K., Walkowicz, L., Willman, B., Yoachim, P., Anderson, S., Antilogus, P., Angus, R., Arcavi, I., Awan, H., Biswas, R., Bell, K. J., Bennett, D., Britt, C., Buzasi, D., Casetti-Dinescu, D. I., Chomiuk, L., Claver, C., Cook, K., Davenport, J., Debattista, V., Digel, S., Doctor, Z., Firth, R. E., Foley, R., Fong, W.-f., Galbany, L., Giampapa, M., Gizis, J. E., Graham, M. L., Grillmair, C., Gris, P., Haiman, Z., Hartigan, P., Hawley, S., Hlozek, R., Jha, S. W., Johns-Krull, C., Kanbur, S., Kalogera, V., Kashyap, V., Kasliwal, V., Kessler, R., Kim, A., Kurczynski, P., Lahav, O., Liu, M. C., Malz, A., Margutti, R., Matheson, T., McEwen, J. D., McGehee, P., Meibom, S., Meyers, J., Monet, D., Neilsen, E., Newman, J., O'Dowd, M., Peiris, H. V., Penny, M. T., Peters, C., Poleski, R., Ponder, K., Richards, G., Rho, J., Rubin, D., Schmidt, S., Schuhmann, R. L., Shporer, A., Slater, C., Smith, N., Soares-Santos, M., Stassun, K., Strader, J., Strauss, M., Street, R., Stubbs, C., Sullivan, M., Szkody, P., Trimble, V., Tyson, T., de Val-Borro, M., Valenti, S., Wagoner, R., Wood-Vasey, W. M., and Zauderer, B. A. (2017). Science-Driven Optimization of the LSST Observing Strategy. *ArXiv e-prints*.

Lyke, B. W., Higley, A. N., McLane, J. N., Schurhammer, D. P., Myers, A. D., Ross, A. J., Dawson, K., Chabanier, S., Martini, P., Busca, N. G., Mas des Bourboux, H. d., Salvato, M., Streblyanska, A., Zarrouk, P., Burtin, E., Anderson, S. F., Bautista, J., Bizyaev, D., Brandt, W. N., Brinkmann, J., Brownstein, J. R., Comparat, J., Green, P., de la Macorra, A., Muñoz Gutiérrez, A., Hou, J., Newman, J. A., Palanque-Delabrouille, N., Pâris, I., Percival, W. J., Petitjean, P., Rich, J., Rossi, G., Schneider, D. P., Smith, A., Vivek, M., and Weaver, B. A. (2020). The Sloan Digital Sky Survey Quasar Catalog: Sixteenth Data Release. *ApJs*, 250(1):8.

Lyth, D. H., Ungarelli, C., and Wands, D. (2003). Primordial density perturbation in the curvaton scenario. *Physical Review D*, 67(2):023503.

Maldacena, J. (2003a). Non-gaussian features of primordial fluctuations in single field inflationary models. *Journal of High Energy Physics*, 2003(05):013.

Maldacena, J. (2003b). Non-gaussian features of primordial fluctuations in single field inflationary models. *Journal of High Energy Physics*, 2003(05):013–013.

Matarrese, S. and Verde, L. (2008). The effect of primordial non-gaussianity on halo bias. *The Astrophysical Journal Letters*, 677(2):L77.

Matarrese, S., Verde, L., and Jimenez, R. (2000). The abundance of high-redshift objects as a probe of non-gaussian initial conditions. *The Astrophysical Journal*, 541(1):10.

Mather, J. C., Cheng, E. S., Eplee, Jr., R. E., Isaacman, R. B., Meyer, S. S., Shafer, R. A., Weiss, R., Wright, E. L., Bennett, C. L., Boggess, N. W., Dwek, E., Gulkis, S., Hauser, M. G., Janssen, M., Kelsall, T., Lubin, P. M., Moseley, Jr., S. H., Murdock, T. L., Silverberg, R. F., Smoot, G. F., and Wilkinson, D. T. (1990). A preliminary measurement of the cosmic microwave background spectrum by the Cosmic Background Explorer (COBE) satellite. *APJL*, 354:L37–L40.

Mehta, V., Scarlata, C., Colbert, J. W., Dai, Y. S., Dressler, A., Henry, A., Malkan, M., Rafelski, M., Siana, B., Teplitz, H. I., Bagley, M., Beck, M., Ross, N. R., Rutkowski, M., and Wang, Y. (2015). Predicting the Redshift 2 Hα Luminosity Function Using [OIII] Emission Line Galaxies. *ApJ*, 811(2):141.

Merz, G., Rezaie, M., Seo, H.-J., Neveux, R., Ross, A. J., Beutler, F., Percival, W. J., Mueller, E., Gil-Marín, H., Rossi, G., et al. (2021). The clustering of the sdss-iv extended baryon oscillation spectroscopic survey quasar sample: Testing observational systematics on the baryon acoustic oscillation measurement. *arXiv preprint arXiv:2105.10463*.

Mohammad, F. G., Percival, W. J., Seo, H.-J., Chapman, M. J., Bianchi, D., Ross, A. J., Zhao, C., Lang, D., Bautista, J., Brinkmann, J., Brownstein, J. R., Burtin, E., Chuang, C.-H., Dawson, K. S., de la Torre, S., de Mattia, A., Eftekharzadeh, S., Fromenteau, S., Gil-Marín, H., Hou, J., Mueller, E.-M., Neveux, R., Paviot, R., Raichoor, A., Rossi, G., Schneider, D. P., Tamone, A., Tinker, J. L., Tojeiro, R., Vargas Magaña, M., and Zhao, G.-B. (2020). The completed SDSS-IV extended baryon oscillation spectroscopic survey: pairwise-inverse probability and angular correction for fibre collisions in clustering measurements. *MNRAS*, 498(1):128–143.

Montufar, G. F., Pascanu, R., Cho, K., and Bengio, Y. (2014). On the number of linear regions of deep neural networks. In *Advances in neural information processing systems*, pages 2924–2932.

Mueller, E.-M. and et al. in prep. (2020). Xxxxxxxxx. *XXXX*, XXX(X):XXX–XXX.

Mueller, E.-M., Percival, W. J., and Ruggeri, R. (2019). Optimizing primordial non-gaussianity measurements from galaxy surveys. *Monthly Notices of the Royal Astronomical Society*, 485(3):4160–4166.

Mukhanov, V. F. and Chibisov, G. V. (1981). Quantum fluctuations and a nonsingular universe. *Soviet Journal of Experimental and Theoretical Physics Letters*, 33:532.

Mukhanov, V. F., Feldman, H. A., and Brandenberger, R. H. (1992). Theory of cosmological perturbations. *Physics Reports*, 215(5-6):203–333.

Myers, A. D., Brunner, R. J., Richards, G. T., Nichol, R. C., Schneider, D. P., and Bahcall, N. A. (2007). Clustering analyses of 300,000 photometrically classified quasars. ii. the excess on very small scales. *The Astrophysical Journal*, 658(1):99.

Myers, A. D., Palanque-Delabrouille, N., Prakash, A., Pâris, I., Yeche, C., Dawson, K. S., Bovy, J., Lang, D., Schlegel, D. J., Newman, J. A., et al. (2015). The sdss-iv extended baryon oscillation spectroscopic survey: Quasar target selection. *The Astrophysical Journal Supplement Series*, 221(2):27.

Nair, V. and Hinton, G. E. (2010). Rectified linear units improve restricted boltzmann machines. In *Proceedings of the 27th international conference on machine learning (ICML-10)*, pages 807–814.

Neveux, R., Burtin, E., de Mattia, A., Smith, A., Ross, A. J., Hou, J., Bautista, J., Brinkmann, J., Chuang, C.-H., Dawson, K. S., Gil-Marín, H., Lyke, B. W., de la Macorra, A., du Mas des Bourboux, H., Mohammad, F. G., Müller, E.-M., Myers, A. D., Newman, J. A., Percival, W. J., Rossi, G., Schneider, D., Vivek, M., Zarrouk, P., Zhao, C., and Zhao, G.-B. (2020). The completed SDSS-IV extended Baryon Oscillation Spectroscopic Survey: BAO and RSD measurements from the anisotropic power spectrum of the quasar sample between redshift 0.8 and 2.2. *MNRAS*, 499(1):210–229.

Padmanabhan, N., Schlegel, D. J., Seljak, U., Makarov, A., Bahcall, N. A., Blanton, M. R., Brinkmann, J., Eisenstein, D. J., Finkbeiner, D. P., Gunn, J. E., Hogg, D. W., Ivezić, Ž., Knapp, G. R., Loveday, J., Lupton, R. H., Nichol, R. C., Schneider, D. P., Strauss, M. A., Tegmark, M., and York, D. G. (2007). The clustering of luminous red galaxies in the Sloan Digital Sky Survey imaging data. *MNRAS*, 378(3):852–872.

Padmanabhan, N. and White, M. (2008). Constraining anisotropic baryon oscillations. *Physical Review D*, 77(12):123540.

Palazzo, A. (2013). Phenomenology of light sterile neutrinos: a brief review. *Modern Physics Letters A*, 28(07):1330004.

Partridge, R. B. (2007). *3K: the cosmic microwave background radiation*, volume 25. Cambridge University Press.

Patrignani, C., Weinberg, D., Woody, C., Chivukula, R., Buchmueller, O., Kuyanov, Y. V., Blucher, E., Willocq, S., Höcker, A., Lippmann, C., et al. (2016). Review of particle physics. *Chin. Phys.*, 40:100001.

Peebles, P. (1973). Statistical analysis of catalogs of extragalactic objects. i. theory. *The Astrophysical Journal*, 185:413–440.

Peebles, P. J. E. and Yu, J. T. (1970). Primeval Adiabatic Perturbation in an Expanding Universe. *ApJ*, 162:815.

Penzias, A. A. and Wilson, R. W. (1965). A measurement of excess antenna temperature at 4080 mc/s. *The Astrophysical Journal*, 142:419–421.

Percival, W. J., Reid, B. A., Eisenstein, D. J., Bahcall, N. A., Budavari, T., Frieman, J. A., Fukugita, M., Gunn, J. E., Ivezić, Ž., Knapp, G. R., et al. (2010). Baryon acoustic oscillations in the sloan digital sky survey data release 7 galaxy sample. *Monthly Notices of the Royal Astronomical Society*, 401(4):2148–2168.

Perlmutter, S., Aldering, G., Goldhaber, G., Knop, R., Nugent, P., Castro, P., Deustua, S., Fabbro, S., Goobar, A., Groom, D., et al. (1999). Measurements of ω and λ from 42 high-redshift supernovae. *The Astrophysical Journal*, 517(2):565.

Ponthieu, N., Grain, J., and Lagache, G. (2011). POKER: estimating the power spectrum of diffuse emission with complex masks and at high angular resolution. *AAP*, 535:A90.

Prakash, A., Licquia, T. C., Newman, J. A., Ross, A. J., Myers, A. D., Dawson, K. S., Kneib, J.-P., Percival, W. J., Bautista, J. E., Comparat, J., et al. (2016). The sdss-iv extended baryon oscillation spectroscopic survey: Luminous red galaxy target selection. *The Astrophysical Journal Supplement Series*, 224(2):34.

Pullen, A. R. and Hirata, C. M. (2013). Systematic effects in large-scale angular power spectra of photometric quasars and implications for constraining primordial non-gaussianity. *Publications of the Astronomical Society of the Pacific*, 125(928):705.

Raichoor, A., Comparat, J., Delubac, T., Kneib, J.-P., Yèche, C., Dawson, K., Percival, W., Dey, A., Lang, D., Schlegel, D., et al. (2017). The sdss-iv extended baryon oscillation spectroscopic survey: final emission line galaxy target selection. *Monthly Notices of the Royal Astronomical Society*, 471(4):3955–3973.

Raichoor, A., Comparat, J., Delubac, T., Kneib, J.-P., Yèche, C., Dawson, K. S., Percival, W. J., Dey, A., Lang, D., Schlegel, D. J., Gorgoni, C., Bautista, J., Brownstein, J. R., Mariappan, V., Seo, H.-J., Tinker, J. L., Ross, A. J., Wang, Y., Zhao, G.-B., Moustakas, J., Palanque-Delabrouille, N., Jullo, E., Newmann, J. A., Prada, F., and Zhu, G. B. (2017). The SDSS-IV extended Baryon Oscillation Spectroscopic Survey: final emission line galaxy target selection. *MNRAS*, 471:3955–3973.

Raichoor, A., de Mattia, A., Ross, A. J., Zhao, C., Alam, S., Avila, S., Bautista, J., Brinkmann, J., Brownstein, J. R., Burtin, E., Chapman, M. J., Chuang, C.-H., Comparat, J., Dawson, K. S., Dey, A., du Mas des Bourboux, H., Elvin-Poole, J., Gonzalez-Perez, V., Gorgoni, C., Kneib, J.-P., Kong, H., Lang, D., Moustakas, J., Myers, A. D., Müller, E.-M., Nadathur, S., Newman, J. A., Percival, W. J., Rezaie, M., Rossi, G., Ruhlmann-Kleider, V., Schlegel, D. J., Schneider, D. P., Seo, H.-J., Tamone, A., Tinker, J. L., Tojeiro, R., Vivek, M., Yèche, C., and Zhao, G.-B. (2020). The completed SDSS-IV extended Baryon Oscillation Spectroscopic Survey: large-scale structure catalogues and measurement of the isotropic BAO between redshift 0.6 and 1.1 for the Emission Line Galaxy Sample. *MNRAS*, 500(3):3254–3274.

Ramaswamy, S., Tamayo, P., Rifkin, R., Mukherjee, S., Yeang, C.-H., Angelo, M., Ladd, C., Reich, M., Latulippe, E., Mesirov, J. P., et al. (2001). Multiclass cancer diagnosis using tumor gene expression signatures. *Proceedings of the National Academy of Sciences*, 98(26):15149–15154.

Rees, M. J. and Sciama, D. W. (1968). Large-scale density inhomogeneities in the universe. *Nature*, 217(5128):511–516.

Reid, B. A., Verde, L., Dolag, K., Matarrese, S., and Moscardini, L. (2010). Non-Gaussian halo assembly bias. *JCAP*, 2010(7):013.

Rezaie, M., Seo, H.-J., Ross, A. J., and Bunescu, R. C. (2020). Improving galaxy clustering measurements with deep learning: analysis of the DECaLS DR7 data. *MNRAS*, 495(2):1613–1640.

Rhodes, J., Allen, S., Benson, B. A., Chang, T., de Putter, R., Dodelson, S., Dore, O., Honscheid, K., Linder, E., Menard, B., Newman, J., Nord, B., Rozo, E., Vallinotto, A., and Weinberg, D. (2013). Exploiting Cross Correlations and Joint Analyses. *arXiv e-prints*, page arXiv:1309.5388.

Richards, G. T., Myers, A. D., Gray, A. G., Riegel, R. N., Nichol, R. C., Brunner, R. J., Szalay, A. S., Schneider, D. P., and Anderson, S. F. (2008). Efficient photometric selection of quasars from the sloan digital sky survey. ii. 1, 000, 000 quasars from data release 6. *The Astrophysical Journal Supplement Series*, 180(1):67.

Riess, A. G., Filippenko, A. V., Challis, P., Clocchiatti, A., Diercks, A., Garnavich, P. M., Gilliland, R. L., Hogan, C. J., Jha, S., Kirshner, R. P., et al. (1998). Observational evidence from supernovae for an accelerating universe and a cosmological constant. *The Astronomical Journal*, 116(3):1009.

Rolnick, D. and Tegmark, M. (2017). The power of deeper networks for expressing natural functions. *arXiv preprint arXiv:1705.05502*.

Ross, A. J., Bautista, J., Tojeiro, R., Alam, S., Bailey, S., Burtin, E., Comparat, J., Dawson, K. S., de Mattia, A., du Mas des Bourboux, H., Gil-Marín, H., Hou, J., Kong, H., Lyke, B. W., Mohammad, F. G., Moustakas, J., Mueller, E.-M., Myers, A. D., Percival, W. J., Raichoor, A., Rezaie, M., Seo, H.-J., Smith, A., Tinker, J. L., Zarrouk, P., Zhao, C., Zhao, G.-B., Bizyaev, D., Brinkmann, J., Brownstein, J. R., Rosell, A. C., Chabanier, S., Choi, P. D., Chuang, C.-H., Cruz-Gonzalez, I., de la Macorra, A., de la Torre, S., Escoffier, S., Fromenteau, S., Higley, A., Jullo, E., Kneib, J.-P., McLane, J. N., Muñoz-Gutiérrez, A., Neveux, R., Newman, J. A., Nitschelm, C., Palanque-Delabrouille, N., Paviot, R., Pullen, A. R., Rossi, G., Ruhlmann-Kleider, V., Schneider, D. P., Magaña,

M. V., Vivek, M., and Zhang, Y. (2020). The Completed SDSS-IV extended Baryon Oscillation Spectroscopic Survey: Large-scale structure catalogues for cosmological analysis. *MNRAS*, 498(2):2354–2371.

Ross, A. J., Beutler, F., Chuang, C.-H., Pellejero-Ibanez, M., Seo, H.-J., Vargas-Magaña, M., Cuesta, A. J., Percival, W. J., Burden, A., Sánchez, A. G., Grieb, J. N., Reid, B., Brownstein, J. R., Dawson, K. S., Eisenstein, D. J., Ho, S., Kitaura, F.-S., Nichol, R. C., Olmstead, M. D., Prada, F., Rodríguez-Torres, S. A., Saito, S., Salazar-Albornoz, S., Schneider, D. P., Thomas, D., Tinker, J., Tojeiro, R., Wang, Y., White, M., and Zhao, G.-b. (2017a). The clustering of galaxies in the completed SDSS-III Baryon Oscillation Spectroscopic Survey: observational systematics and baryon acoustic oscillations in the correlation function. *MNRAS*, 464:1168–1191.

Ross, A. J., Beutler, F., Chuang, C.-H., Pellejero-Ibanez, M., Seo, H.-J., Vargas-Magaña, M., Cuesta, A. J., Percival, W. J., Burden, A., Sánchez, A. G., Grieb, J. N., Reid, B., Brownstein, J. R., Dawson, K. S., Eisenstein, D. J., Ho, S., Kitaura, F.-S., Nichol, R. C., Olmstead, M. D., Prada, F., Rodríguez-Torres, S. A., Saito, S., Salazar-Albornoz, S., Schneider, D. P., Thomas, D., Tinker, J., Tojeiro, R., Wang, Y., White, M., and Zhao, G.-b. (2017b). The clustering of galaxies in the completed SDSS-III Baryon Oscillation Spectroscopic Survey: observational systematics and baryon acoustic oscillations in the correlation function. *MNRAS*, 464(1):1168–1191.

Ross, A. J., Brunner, R. J., and Myers, A. D. (2007). Higher order angular galaxy correlations in the sdss: Redshift and color dependence of nonlinear bias. *The Astrophysical Journal*, 665(1):67.

Ross, A. J., Ho, S., Cuesta, A. J., Tojeiro, R., Percival, W. J., Wake, D., Masters, K. L., Nichol, R. C., Myers, A. D., de Simoni, F., et al. (2011). Ameliorating systematic uncertainties in the angular clustering of galaxies: a study using the sdss-iii. *Monthly Notices of the Royal Astronomical Society*, 417(2):1350–1373.

Ross, A. J., Percival, W. J., Carnero, A., Zhao, G.-b., Manera, M., Raccanelli, A., Aubourg, E., Bizyaev, D., Brewington, H., Brinkmann, J., Brownstein, J. R., Cuesta, A. J., da Costa, L. A. N., Eisenstein, D. J., Ebelke, G., Guo, H., Hamilton, J.-C., Magaña, M. V., Malanushenko, E., Malanushenko, V., Maraston, C., Montesano, F., Nichol, R. C., Oravetz, D., Pan, K., Prada, F., Sánchez, A. G., Samushia, L., Schlegel, D. J., Schneider, D. P., Seo, H.-J., Sheldon, A., Simmons, A., Snedden, S., Swanson, M. E. C., Thomas, D., Tinker, J. L., Tojeiro, R., and Zehavi, I. (2013). The clustering of galaxies in the SDSS-III DR9 Baryon Oscillation Spectroscopic Survey: constraints on primordial non-Gaussianity. *MNRAS*, 428(2):1116–1127.

Ross, A. J., Percival, W. J., Carnero, A., Zhao, G.-b., Manera, M., Raccanelli, A., Aubourg, E., Bizyaev, D., Brewington, H., Brinkmann, J., et al. (2013). The clustering of galaxies in the sdss-iii dr9 baryon oscillation spectroscopic survey: constraints on primordial non-gaussianity. *Monthly Notices of the Royal Astronomical Society*, 428(2):1116–1127.

Ross, A. J., Percival, W. J., Sánchez, A. G., Samushia, L., Ho, S., Kazin, E., Manera, M., Reid, B., White, M., Tojeiro, R., McBride, C. K., Xu, X., Wake, D. A., Strauss, M. A., Montesano, F., Swanson, M. E. C., Bailey, S., Bolton, A. S., Dorta, A. M., Eisenstein, D. J., Guo, H., Hamilton, J.-C., Nichol, R. C., Padmanabhan, N., Prada, F., Schlegel, D. J., Magaña, M. V., Zehavi, I., Blanton, M., Bizyaev, D., Brewington, H., Cuesta, A. J., Malanushenko, E., Malanushenko, V., Oravetz, D., Parejko, J., Pan, K., Schneider, D. P., Shelden, A., Simmons, A., Snedden, S., and Zhao, G.-b. (2012a). The clustering of galaxies in the SDSS-III Baryon Oscillation Spectroscopic Survey: analysis of potential systematics. *MNRAS*, 424:564–590.

Ross, A. J., Percival, W. J., Sánchez, A. G., Samushia, L., Ho, S., Kazin, E., Manera, M., Reid, B., White, M., Tojeiro, R., McBride, C. K., Xu, X., Wake, D. A., Strauss, M. A., Montesano, F., Swanson, M. E. C., Bailey, S., Bolton, A. S., Dorta, A. M., Eisenstein, D. J., Guo, H., Hamilton, J.-C., Nichol, R. C., Padmanabhan, N., Prada, F., Schlegel, D. J., Magaña, M. V., Zehavi, I., Blanton, M., Bizyaev, D., Brewington, H., Cuesta, A. J., Malanushenko, E., Malanushenko, V., Oravetz, D., Parejko, J., Pan, K., Schneider, D. P., Shelden, A., Simmons, A., Snedden, S., and Zhao, G.-b. (2012b). The clustering of galaxies in the SDSS-III Baryon Oscillation Spectroscopic Survey: analysis of potential systematics. *MNRAS*, 424(1):564–590.

Ruder, S. (2016). An overview of gradient descent optimization algorithms. *arXiv e-prints*, page arXiv:1609.04747.

Ruder, S. (2016). An overview of gradient descent optimization algorithms. *arXiv preprint arXiv:1609.04747*.

Ruggeri, R., Percival, W. J., Gil-Marín, H., Zhu, F., Zhao, G.-B., and Wang, Y. (2017). Optimal redshift weighting for redshift-space distortions. *Monthly Notices of the Royal Astronomical Society*, 464(3):2698.

Rybicki, G. B. and Press, W. H. (1992). Interpolation, realization, and reconstruction of noisy, irregularly sampled data. *The Astrophysical Journal*, 398:169–176.

Sachs, R. K. and Wolfe, A. M. (1967). Perturbations of a cosmological model and angular variations of the microwave background. *The Astrophysical Journal*, 147(1):73–90.

Salopek, D., Bond, J., and Bardeen, J. M. (1989). Designing density fluctuation spectra in inflation. *Physical Review D*, 40(6):1753.

Sánchez, A. G., Baugh, C. M., and Angulo, R. (2008). What is the best way to measure baryonic acoustic oscillations? *Monthly Notices of the Royal Astronomical Society*, 390(4):1470–1490.

Schlafly, E., Finkbeiner, D., Jurić, M., Magnier, E., Burgett, W., Chambers, K., Grav, T., Hodapp, K., Kaiser, N., Kudritzki, R.-P., et al. (2012). Photometric calibration of the first 1.5 years of the pan-starrs1 survey. *The Astrophysical Journal*, 756(2):158.

Schlafly, E. F. and Finkbeiner, D. P. (2011). Measuring reddening with sloan digital sky survey stellar spectra and recalibrating sfd. *The Astrophysical Journal*, 737(2):103.

Schlegel, D. J., Finkbeiner, D. P., and Davis, M. (1998). Maps of dust infrared emission for use in estimation of reddening and cosmic microwave background radiation foregrounds. *The Astrophysical Journal*, 500(2):525.

Schmidt, B. P., Suntzeff, N. B., Phillips, M. M., Schommer, R. A., Clocchiatti, A., Kirshner, R. P., Garnavich, P., Challis, P., Leibundgut, B., Spyromilio, J., et al. (1998). The high-z supernova search: measuring cosmic deceleration and global curvature of the universe using type ia supernovae. *The Astrophysical Journal*, 507(1):46.

Scoccimarro, R. (2015). Fast estimators for redshift-space clustering. *PRD*, 92(8):083532.

Scoccimarro, R., Sefusatti, E., and Zaldarriaga, M. (2004). Probing primordial non-gaussianity with large-scale structure. *Physical Review D*, 69(10):103513.

Scranton, R., Johnston, D., Dodelson, S., Frieman, J. A., Connolly, A., Eisenstein, D. J., Gunn, J. E., Hui, L., Jain, B., Kent, S., et al. (2002). Analysis of systematic effects and statistical uncertainties in angular clustering of galaxies from early sloan digital sky survey data. *The Astrophysical Journal*, 579(1):48.

Sefusatti, E., Crocce, M., Scoccimarro, R., and Couchman, H. M. P. (2016). Accurate estimators of correlation functions in Fourier space. *MNRAS*, 460(4):3624–3636.

Seo, H.-J. and Eisenstein, D. J. (2003). Probing Dark Energy with Baryonic Acoustic Oscillations from Future Large Galaxy Redshift Surveys. *APJ*, 598:720–740.

Seo, H.-J. and Eisenstein, D. J. (2003). Probing dark energy with baryonic acoustic oscillations from future large galaxy redshift surveys. *The Astrophysical Journal*, 598(2):720.

Silk, J. (1968). Cosmic Black-Body Radiation and Galaxy Formation. *ApJ*, 151:459.

Slosar, A., Hirata, C., Seljak, U., Ho, S., and Padmanabhan, N. (2008). Constraints on local primordial non-gaussianity from large scale structure. *Journal of Cosmology and Astroparticle Physics*, 2008(08):031.

Slosar, A., Seljak, U., and Makarov, A. (2004). Exact likelihood evaluations and foreground marginalization in low resolution wmap data. *Physical Review D*, 69(12):123003.

Smee, S. A., Gunn, J. E., Uomoto, A., Roe, N., Schlegel, D., Rockosi, C. M., Carr, M. A., Leger, F., Dawson, K. S., Olmstead, M. D., et al. (2013). The multi-object, fiber-fed spectrographs for the sloan digital sky survey and the baryon oscillation spectroscopic survey. *The Astronomical Journal*, 146(2):32.

Smith, A., Burtin, E., Hou, J., Neveux, R., Ross, A. J., Alam, S., Brinkmann, J., Dawson, K. S., Habib, S., Heitmann, K., Kneib, J.-P., Lyke, B. W., du Mas des Bourboux, H., Mueller, E.-M., Myers, A. D., Percival, W. J., Rossi, G., Schneider, D. P., Zarrouk, P., and Zhao, G.-B. (2020). The completed SDSS-IV extended Baryon Oscillation Spectroscopic Survey: N-body mock challenge for the quasar sample. *MNRAS*, 499(1):269–291.

Smith, L. N. (2015). Cyclical Learning Rates for Training Neural Networks. *arXiv e-prints*, page arXiv:1506.01186.

Smoot, G. F., Bennett, C. L., Kogut, A., Wright, E. L., Aymon, J., Boggess, N. W., Cheng, E. S., de Amici, G., Gulkis, S., Hauser, M. G., Hinshaw, G., Jackson, P. D., Janssen, M., Kaita, E., Kelsall, T., Keegstra, P., Lineweaver, C., Loewenstein, K., Lubin, P., Mather, J., Meyer, S. S., Moseley, S. H., Murdock, T., Rokke, L., Silverberg, R. F., Tenorio, L., Weiss, R., and Wilkinson, D. T. (1992). Structure in the COBE differential microwave radiometer first-year maps. *APJL*, 396:L1–L5.

Spergel, D. N., Verde, L., Peiris, H. V., Komatsu, E., Nolta, M., Bennett, C., Halpern, M., Hinshaw, G., Jarosik, N., Kogut, A., et al. (2003). First-year wilkinson microwave anisotropy probe (wmap)* observations: determination of cosmological parameters. *The Astrophysical Journal Supplement Series*, 148(1):175.

Starobinsky, A. A. (1982). Dynamics of phase transition in the new inflationary universe scenario and generation of perturbations. *Physics Letters B*, 117(3-4):175–178.

Suchyta, E., Huff, E. M., Aleksić, J., Melchior, P., Jouvel, S., MacCrann, N., Ross, A. J., Crocce, M., Gaztanaga, E., Honscheid, K., Leistedt, B., Peiris, H. V., Rykoff, E. S., Sheldon, E., Abbott, T., Abdalla, F. B., Allam, S., Banerji, M., Benoit-Lévy, A., Bertin, E., Brooks, D., Burke, D. L., Carnero Rosell, A., Carrasco Kind, M., Carretero, J., Cunha, C. E., D'Andrea, C. B., da Costa, L. N., DePoy, D. L., Desai, S., Diehl, H. T., Dietrich, J. P., Doel, P., Eifler, T. F., Estrada, J., Evrard, A. E., Flaugher, B., Fosalba, P., Frieman, J., Gerdes, D. W., Gruen, D., Gruendl, R. A., James, D. J., Jarvis, M., Kuehn, K., Kuropatkin, N., Lahav, O., Lima, M., Maia, M. A. G., March, M., Marshall, J. L., Miller, C. J., Miquel, R., Neilsen, E., Nichol, R. C., Nord, B., Ogando, R., Percival, W. J., Reil, K., Roodman, A., Sako, M., Sanchez, E., Scarpine, V., Sevilla-Noarbe, I., Smith, R. C., Soares-Santos, M., Sobreira, F., Swanson, M. E. C., Tarle, G., Thaler, J., Thomas, D., Vikram, V., Walker, A. R., Wechsler, R. H., Zhang, Y., and DES Collaboration (2016). No galaxy left behind: accurate measurements with the faintest objects in the Dark Energy Survey. *MNRAS*, 457:786–808.

Szapudi, I., Prunet, S., Pogosyan, D., Szalay, A. S., and Bond, J. R. (2001). Fast Cosmic Microwave Background Analyses via Correlation Functions. *ApJl*, 548(2):L115–L118.

Tamura, S. and Tateishi, M. (1997). Capabilities of a four-layered feedforward neural network: four layers versus three. *IEEE Transactions on Neural Networks*, 8(2):251–255.

Taruya, A., Koyama, K., and Matsubara, T. (2008). Signature of primordial non-gaussianity on the matter power spectrum. *Physical Review D*, 78(12):123534.

Tegmark, M. (1997). How to measure cmb power spectra without losing information. *Physical Review D*, 55(10):5895.

Tegmark, M. (1997). How to measure CMB power spectra without losing information. *PRD*, 55(10):5895–5907.

Tegmark, M., Eisenstein, D. J., Strauss, M. A., Weinberg, D. H., Blanton, M. R., Frieman, J. A., Fukugita, M., Gunn, J. E., Hamilton, A. J. S., Knapp, G. R., Nichol, R. C., Ostriker, J. P., Padmanabhan, N., Percival, W. J., Schlegel, D. J., Schneider, D. P., Scoccimarro, R., Seljak, U., Seo, H.-J., Swanson, M., Szalay, A. S., Vogeley, M. S., Yoo, J., Zehavi, I., Abazajian, K., Anderson, S. F., Annis, J., Bahcall, N. A., Bassett, B., Berlind, A., Brinkmann, J., Budavari, T., Castander, F., Connolly, A., Csabai, I., Doi, M., Finkbeiner, D. P., Gillespie, B., Glazebrook, K., Hennessy, G. S., Hogg, D. W., Ivezić, Ž., Jain, B., Johnston, D., Kent, S., Lamb, D. Q., Lee, B. C., Lin, H., Loveday, J., Lupton, R. H., Munn, J. A., Pan, K., Park, C., Peoples, J., Pier, J. R., Pope, A., Richmond, M., Rockosi, C., Scranton, R., Sheth, R. K., Stebbins, A., Stoughton, C., Szapudi, I., Tucker, D. L., vanden Berk, D. E., Yanny, B., and York, D. G. (2006). Cosmological constraints from the SDSS luminous red galaxies. *PRD*, 74(12):123507.

Tegmark, M., Hamilton, A. J., Strauss, M. A., Vogeley, M. S., and Szalay, A. S. (1998). Measuring the galaxy power spectrum with future redshift surveys. *The Astrophysical Journal*, 499(2):555.

Tegmark, M., Strauss, M. A., Blanton, M. R., Abazajian, K., Dodelson, S., Sandvik, H., Wang, X., Weinberg, D. H., Zehavi, I., Bahcall, N. A., et al. (2004). Cosmological parameters from sdss and wmap. *Physical Review D*, 69(10):103501.

Thomas, S. A., Abdalla, F. B., and Lahav, O. (2011a). The angular power spectra of photometric sloan digital sky survey luminous red galaxies. *Monthly Notices of the Royal Astronomical Society*, 412(3):1669–1685.

Thomas, S. A., Abdalla, F. B., and Lahav, O. (2011b). Excess clustering on large scales in the megaz dr7 photometric redshift survey. *Physical review letters*, 106(24):241301.

Verde, L., Wang, L., Heavens, A. F., and Kamionkowski, M. (2000). Large-scale structure, the cosmic microwave background and primordial non-gaussianity. *Monthly Notices of the Royal Astronomical Society*, 313(1):141–147.

Wagoner, E. L., Rozo, E., Fang, X., Crocce, M., Elvin-Poole, J., and Weaverdyck, N. (2020). Linear Systematics Mitigation in Galaxy Clustering in the Dark Energy Survey Year 1 Data. *arXiv e-prints*, page arXiv:2009.10854.

Weaverdyck, N. and Huterer, D. (2020). Mitigating contamination in LSS surveys: a comparison of methods. *arXiv e-prints*, page arXiv:2007.14499.

Weinberg, D. H., Mortonson, M. J., Eisenstein, D. J., Hirata, C., Riess, A. G., and Rozo, E. (2013). Observational probes of cosmic acceleration. *Physics reports*, 530(2):87–255.

Welch, B. L. (1947). The generalization ofstudent's' problem when several different population variances are involved. *Biometrika*, 34(1/2):28–35.

White, M., Tinker, J. L., and McBride, C. K. (2013). Mock galaxy catalogues using the quick particle mesh method. *Monthly Notices of the Royal Astronomical Society*, 437(3):2594–2606.

Winberg, S. (1972). Gravitation and cosmology. *ed. John Wiley and Sons, New York*, pages 495–464.

Wright, E. L., Eisenhardt, P. R., Mainzer, A. K., Ressler, M. E., Cutri, R. M., Jarrett, T., Kirkpatrick, J. D., Padgett, D., McMillan, R. S., Skrutskie, M., et al. (2010). The wide-field infrared survey explorer (wise): mission description and initial on-orbit performance. *The Astronomical Journal*, 140(6):1868.

Yamamoto, K., Nakamichi, M., Kamino, A., Bassett, B. A., and Nishioka, H. (2006). A Measurement of the Quadrupole Power Spectrum in the Clustering of the 2dF QSO Survey. *PASJ*, 58:93–102.

York, D. G., Adelman, J., Anderson Jr, J. E., Anderson, S. F., Annis, J., Bahcall, N. A., Bakken, J., Barkhouser, R., Bastian, S., Berman, E., et al. (2000). The sloan digital sky survey: Technical summary. *The Astronomical Journal*, 120(3):1579.

Zaldarriaga, M. (2004). Non-gaussianities in models with a varying inflaton decay rate. *Physical Review D*, 69(4):043508.

Zarrouk, P., Rezaie, M., Raichoor, A., Ross, A. J., Alam, S., Blum, R., Brookes, D., Chuang, C.-H., Cole, S., Dawson, K. S., Eisenstein, D. J., Kehoe, R., Landriau, M., Moustakas, J., Myers, A. D., Norberg, P., Percival, W. J., Prada, F., Schubnell, M., Seo, H.-J., Tarlé, G., and Zhao, C. (2021). Baryon Acoustic Oscillations in the projected cross-correlation function between the eBOSS DR16 quasars and photometric galaxies from the DESI Legacy Imaging Surveys. *MNRAS*.

Zehavi, I., Zheng, Z., Weinberg, D. H., Frieman, J. A., Berlind, A. A., Blanton, M. R., Scoccimarro, R., Sheth, R. K., Strauss, M. A., Kayo, I., Suto, Y., Fukugita, M., Nakamura, O., Bahcall, N. A., Brinkmann, J., Gunn, J. E., Hennessy, G. S., Ivezić, Ž., Knapp, G. R., Loveday, J., Meiksin, A., Schlegel, D. J., Schneider, D. P., Szapudi, I., Tegmark, M., Vogeley, M. S., York, D. G., and SDSS Collaboration (2005). The Luminosity and Color Dependence of the Galaxy Correlation Function. *ApJ*, 630(1):1–27.

Zel'Dovich, Y. B. (1970). Gravitational instability: An approximate theory for large density perturbations. *Astronomy and astrophysics*, 5:84–89.

Zhao, C., Chuang, C.-H., Bautista, J., de Mattia, A., Raichoor, A., Ross, A. J., Hou, J., Neveux, R., Tao, C., Burtin, E., Dawson, K. S., de la Torre, S., Gil-Marín, H., Kneib, J.-P., Percival, W. J., Rossi, G., Tamone, A., Tinker, J. L., Zhao, G.-B., Alam, S., and Mueller, E.-M. (2021). The completed SDSS-IV extended Baryon Oscillation Spectroscopic Survey: 1000 multi-tracer mock catalogues with redshift evolution and systematics for galaxies and quasars of the final data release. *MNRAS*, 503(1):1149–1173.

Zhu, F., Padmanabhan, N., and White, M. (2015). Optimal redshift weighting for baryon acoustic oscillations. *Monthly Notices of the Royal Astronomical Society*, 451:236.

Zhu, F., Padmanabhan, N., White, M., Ross, A. J., and Zhao, G. (2016). Redshift Weights for Baryon Acoustic Oscillations : Application to Mock Galaxy Catalogs. *Monthly Notices of the Royal Astronomical Society*, 461(3):2867–2878. arXiv: 1604.01050.

Zou, H., Zhou, X., Fan, X., Zhang, T., Zhou, Z., Nie, J., Peng, X., McGreer, I., Jiang, L., Dey, A., et al. (2017). Project overview of the beijing–arizona sky survey. *Publications of the Astronomical Society of the Pacific*, 129(976):064101.

Zwicky, F. (1933). Die rotverschiebung von extragalaktischen nebeln. *Helvetica Physica Acta*, 6:110–127.

Zwicky, F. (1937). On the masses of nebulae and of clusters of nebulae. *The Astrophysical Journal*, 86:217.

Appendix: Survey Geometry

A.1 Angular Power Spectrum

The observed density field of targets does not cover the full sky, due to the galactic plane obscuration. This means that the pseudo-power spectrum $\hat{C}_\ell$ obtained by the direct Spherical Harmonic Transforms of a partial sky map, differs from the full-sky angular spectrum C_ℓ. However, their ensemble average is related by (Hivon et al., 2002; Ponthieu et al., 2011)

$$< \hat{C}_\ell > = \sum_{\ell'} M_{\ell\ell'} < C_{\ell'} >, \tag{A.1}$$

where $M_{\ell\ell'}$ represents the mode-mode coupling from the partial sky coverage. This is known as the Window Function effect and a proper assessment of this effect is crucial for a robust measurement of the large-scale clustering of galaxies. We follow a similar approach to that of (Szapudi et al., 2001; Chon et al., 2004) to model the window function effect on the theoretical power spectrum C_ℓ rather than correcting the measured pseudo-power spectrum from data. First, we compute the two-point correlation function of the window,

$$RR(\theta) = \sum_{i,j>1} f_{\mathrm{pix,i}} f_{\mathrm{pix,j}} \Theta_{ij}(\theta), \tag{A.2}$$

where $\Theta_{ij}(\theta)$ is one when the pixels i and j are separated by an angle between θ and $\theta + \Delta\theta$, and zero otherwise. Next, we normalize the RR by $\sin(\theta)\Delta\theta$ to account for the area and total number of pairs such that $RR(\theta = 0) = 1$. We fit a polynomial on RR to smooth out the wiggles raised by noise. Then, we multiply the theoretical correlation function $\omega(\theta)$ by the window paircount,

$$\omega^{WC}(\theta) = \omega(\theta)\, RR(\theta), \tag{A.3}$$

and finally use the Gaussian-Quadrature algorithm to transform the window convolved theory correlation function ω^{WC} to C_ℓ^{WC},

$$C_\ell^{WC} = 2\pi \int \omega^{WC}(\theta) P_\ell(\theta) d\theta. \tag{A.4}$$

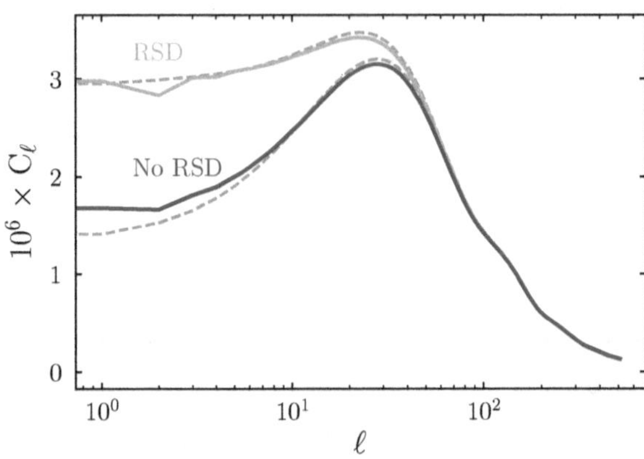

Figure A.1: Window corrected theory C_ℓ for two different models with and without Redshift Space Distortions respectively in orange and blue. The dashed curves show the theoretical models before window convolution. The effect of the window is around 5% in redshift and 20% in real space. The theory with redshift space distortions uses *galaxy bias* = 2 and the surface density $n(z)$ of NGC eBOSS ELG (Tab. 4 of Raichoor et al., 2017) and assuming the fiducial cosmology of Ross et al. (2012a); Ho et al. (2012).

Fig. A.1 shows the DECaLS window effect on two theoretical models of C_ℓ with and without redshift space distortions. The window effect for the model without Redshift Space Distortions (RSD) is around 20% but that for the model with RSD is less than 5% due to the flat power spectrum at the low ell limit. We find a consistent pattern in the mocks; the window effect on the clustering of the mocks is between 5-15%.